Winner of The People's
What the People said

Essential and timely. A great book for us all.

Preserving the earth for future generations of all living things is vital. This book focuses on how to do this legally, how to wake the public to possible solutions. I vote for *Eradicating Ecocide* by Polly Higgins.

A fantastic piece of pursuasive argument. This should be read by everyone in No.10 and the White House.

I vote for *Eradicating Ecocide*.

This is the book that legal systems around the world are waiting for – drawing on its ideas to overturn 500 years of denying nature a legal voice. This should be book of the century.

This book is one of the most important books for solving our most vital problem … it is literally about the survival of life as we know it on our planet! Polly brings us what we so desperately need: finally, an elegant, inspiring, ambitious yet totally possible solution to save ourselves!

Go Polly! xxx

This book is important.

If anyone can help save the planet, Polly's book can.

Not the most optimistic book, but *crucial* for our future wellbeing, or even survival!

If this book fails to win the prize, humanity deserves what it's got in store…

Important ideas and practical application!

Essential reading!

A must read for anyone who cares about having a socially just and liveable tomorrow – Jim

Very good.

An essential book that can help us create a better world.

Voting remains open until midnight, May 31st.

Interesting read!

Polly's book is essential reading to anyone interested in looking after the world and not interested in green wash. Gareth

More comments at the back of the book

Earth is our Business

ALSO BY POLLY HIGGINS
Eradicating Ecocide (2010)

Earth is our Business
changing the rules of the game

POLLY HIGGINS

SHEPHEARD-WALWYN (PUBLISHERS) LTD

First published in 2012 by
Shepheard-Walwyn (Publishers) Ltd
107 Parkway House, Sheen Lane
London SW14 8LS
www.shepheard-walwyn.co.uk

British Library Cataloguing in Publication Data
A catalogue record of this book
is available from the British Library

ISBN-13: 978-0-85683-288-8

Cover design by Tentacle
www.tentacledesign.co.uk
Typeset by Arthouse Publishing Solutions Ltd,
www.arthousepublishing.co.uk
Printed and bound in the USA
by Edwards Brothers Inc.

DEDICATION

TO ALL WHO CARE FOR OUR EARTH
TO ALL WHO BELIEVE THAT EARTH IS OUR BUSINESS AND
TO ALL WHO ARE IN SERVICE TO THE WIDER EARTH
COMMUNITY

Contents

Acknowledgements

I THANK all who are travelling on this journey with me and who have helped me along the way to each destination and viewpoint. Without you, the journey would not have happened. Each and every person who has stopped to show me the way has been like a marker on the road to the summit. It's a rocky and steep ascent, with lots of paths going off in other directions. Some have been explored, others have been missed. I give thanks to you all.

Introduction

'CRIMES Against Peace' is a generic term used to describe four existing crimes that are deemed to be so abhorrent that they have been identified as international crimes against peace. These apply to humanity as a whole, regardless of whether or not your country has put in place the laws to prevent them. They are crimes which cause the diminution of our right to peaceful enjoyment of life. Globally they are considered the worst of all crimes; the four are genocide, crimes against humanity, war crimes and crimes of aggression.

They are predominantly crimes which protect our human right to life. One of them, however, identifies environmental destruction as a crime, and then only during war-time, not peace-time. Thus, we have a missing 5th crime against peace: ecocide – the environmental equivalent of genocide. This book expands on my first book, *Eradicating Ecocide*, which sets out the legal and moral premise for making ecocide a crime. Here I examine the criminal aspects of the Law of Ecocide in full detail with a sample indictment and Ecocide Act which was used as the basis for a mock trial in the UK Supreme Court on 30 September 2011. Both were tested and the outcome is an Act that is ready to implement, when ecocide is made the 5th crime against peace.

The crime of ecocide is a natural evolution of law: the Ecocide Act, set out in Appendix 2, is not radical in its remit. On the contrary, it is part of an evolution of legislation dealing with the impact of pollution and the principle of superior responsibility. In the eyes of the law, creating the crime of ecocide is not about closing the door to evil. It is in fact about protecting a higher value: the sacredness of life, all life.

Those who are *prima facie* guilty of committing ecocide are not in themselves evil – many companies have bought into the norm that it is collateral damage to destroy the earth whilst serving humanity. There is rarely wilful intent where companies are looking to help satisfy human needs, such as energy. Rather it is a blindness that prevents many from facing the truth that human needs can be well served without diminishing the earth's capacity to support life as we know it.

Genocide, unlike ecocide, was viewed as an incomparable evil. Slavery was viewed as a manifest evil. Both were moments in history when we reached a junction – prohibit and prevent or allow it to continue. Before laws were made prohibiting both genocide and slavery, neither were illegal: in fact both generated profit for many parties. The prohibitions that followed did not mean that economies collapsed. New ones evolved and new ways were found. What was once the norm, became overnight the exception. It was law that shifted societal norms. The law has a powerful force which can shape our world in ways that we can hardly comprehend. It took the holocaust to drive in the new way of thinking that gassing humans was a crime. Prior to that, it wasn't recognised as an international crime, which made it almost impossible for people like Sophie Scholl to stand up and object. She and others in Nazi Germany were fighting against something that had been endorsed by their government and the media as the norm, no matter how unpleasant it was. In so doing, the people were effectively silenced. Without the word for genocide in their vocabulary, it was almost impossible to identify what was a crime. Without it, all remained hidden in the eyes of the world for quite some time.

Genocide was justified on self-interest and collective rationality, obscene though it seems to us today. Now catastrophic corporate rationalism places self-interest and growth as justification for destruction of the environment. Those who are guilty of destroying our planet, rationalise their actions by saying they have the right to make money without taking responsibility for decisions that adversely impact all life as we know it. This is our blindness.

Climate change is just a symptom. Like a cold, we hope we can brave it out until it recedes. But this is one cold that has turned serious, not just for you and me but for the whole of humanity. The problem is we are treating it with thinly disguised placebos in the hope that they will do the trick. Without addressing the source, the symptom has no chance of being cured. Instead the symptom returns time and again, each time worse and increasingly debilitating. In time we become accustomed to the debilitation and accept it. Yet still it gets worse: like a smoker who is hacking and coughing but nonetheless drags deeply on his cigarette, choking in the knowledge that his behaviour is facilitating his own painful death. So too are we continuing to indulge a habit that has no benefit for us either in the short or the long term.

The difference is that this particular malaise is born of our failure to take responsibility for the health and well-being of planet Earth. Our bodies are capable of withstanding much abuse, but our planet has reached such a point of damage that her health is at risk of tipping over the edge into an abyss where humanity can no longer be sustained. We can ignore the reality with which we are faced: death, destruction and loss of species on an unprecedented scale, or we can face the truth and meet the consequences face on.

No-one is calling for this Armageddon to stop; no-one is standing up and refusing to participate. We have all become complicit without questioning the consequences. Those who stand at the helm of their businesses are prevented from doing so by the law as it stands which makes profit the primary obligation, even when it means the end of our world as we know it. Now is the time to establish an over-riding duty of care as our number one priority – one that ensures that the welfare of the people and planet is placed above the corporate duty to make money for shareholders. Business has the potential to be great, to be the solution and not the problem. It will require new laws to make that happen and this book sets out the law that can do just that. The aim of this book is to enable business and governments to take the necessary steps in a different direction from the way we are going.

All existing proposals fail to disrupt the very system that is destroying our world. Of those that have been put on the table, none are enforceable, none are capable of delivering on time and none have proven to be turnkeys. Not one of the proposals will effectively halt dangerous industrial activity: the replacement to the Kyoto Protocol (proposed to come into force in 2020) is voluntary; a Green Fund with no funds and the $100 billion promise will not be provided by the developed countries; REDD (Reducing Emissions from Deforestation and Forest Degradation) has failed to safeguard the people and funding has been postponed until the next decade.

2020, it's too late to wait: a very different route can be taken instead. What is needed is a disruptor to our current trajectory and a law to set a framework for intervention. To rely on existing policies is a miscarriage of justice.

This is a story with two possible endings: one is fertile and abundant with life, the other is arid and speaks of death. We have a choice: to make the leap to the new and leave the old ways behind as distant memories, or follow the current route into the ecocide of the earth. By setting out the legal tools we can use, our choice can be life-affirming and can be a decision which will ensure a positive outlook for many beings. Let's face the challenge head on together.

Part 1

WHERE THE WORLD IS CURRENTLY HEADING

All it takes is for one person to stand up and speak out

Chapter 1

THE LAW OF ECOCIDE

Ecocide is the extensive damage to, destruction of or loss of ecosystem(s) of a given territory, whether by human agency or by other causes, to such an extent that peaceful enjoyment by the inhabitants of that territory has been severely diminished.

A T certain points in history the world changes gear. We abolished slavery, apartheid was outlawed and we criminalised genocide. Each time humanity reached a tipping point; no longer could we justify using blacks as slaves, destroy lives and allow others to determine the outcome of a man's life. We get to a stage that we turn and face the truth, even when it is not a sight we wish to see, we give it a name and we say, 'no more'.

We are now at another point of acceleration; we are poised to move the gear stick up to the next level. We have our foot on the pedal and we are ready to go. But wait. To go to the next level we need new rules. Number one rule is set out below, others are contained within this book. Collectively they make for a safe journey into the unknown. Treat this book as your guide to take with you on your journey, to equip you with the language and the route map to the new world.

Ecocide is 'the extensive damage, destruction to or loss of ecosystems of a given territory, whether by human agency or by other causes, to such an extent that peaceful enjoyment by the inhabitants of that territory has been severely diminished.'

The Law of Ecocide is a law which will change the world. The ramifications for business are huge and the lives of all who live on Earth. It will signal the beginning of business taking full responsibility. Humanity will celebrate the end of a polluting and destructive era. The earth will be given a chance to heal.

Ecocide comes in many forms and is either human-made or caused by catastrophic disaster. Human-made ecocide is corporate-driven activity such as deforestation, pollution dumping, mining. Natural ecocide includes tsunamis, floods, earthquakes, rises in sea-levels – in short any event which causes mass ecosystem collapse.

The Law of Ecocide imposes a superior obligation and a pre-emptive legal duty upon individuals who are in a position of superior responsibility within corporations, banks and governments to prohibit profit, investment and policy which causes or supports ecocide. The crime of ecocide criminalises damage, destruction or loss of ecosystems over a certain size, duration and impact. Make ecocide unlawful and a legal framework of nation-to-nation responsibility can be set up to finance humanitarian and environmental aid for ecocide-affected territories.

CRIME AGAINST PEACE

There are certain principles of universal validity and application that apply to humanity as a whole. They are the principles that underpin the prohibition of certain behaviour, for example apartheid and genocide. Such abuses arise out of value systems based on a lack of regard for human life and are now universally outlawed. The most serious of all have been declared Crimes Against Peace by the United Nations and they apply across the world, superseding all other laws. A value system based on a lack of regard for all life now needs to be universally outlawed as well. Kill our planet and we kill ourselves. Ecocide is death by a thousand cuts: each day the life-source which feeds and nourishes our human life is damaged and destroyed a little more. Restoration of territories which have been subjected to human ecocide is not being undertaken voluntarily and as a result conflict and resource wars are expected to escalate over time.

Creation of the Law of Ecocide will close the door to investment in high-risk ventures which give rise to ecocide. Decision-making will be determined on a value-driven basis premised on intrinsic values, not permit allocations. Protection of the interests of the wider Earth community will then become the over-riding consideration for business, driving innovation in a new direction.

RULES OF THE GAME

That is all that law is – rules of the game of life, rules that we humans have put in place. Law is a constantly evolving field and the rules constantly change, become modified and are expanded. Law has the ability to change the playing field radically, overnight. We can play as if there is no tomorrow, or we can look over the horizon and decide to engage in the new rules before they arrive. Thus, when we do, we have already honed our skills and are ready to move fast in a direction we are already heading in.

Any company which has an eye to the future will want to flow with the times. Our corporate culture is predicated on evaluating what is most likely to happen if business stays the same, not looking to how things can change. Banks are now having closed door conversations with others about restructuring their approach so that a principled system is put in place. They are rethinking the problem through a lens that is placing intrinsic values at the centre.

When the existing system fails to prevent that which it is set up to help, the scales of justice swing out of kilter and the rules of the game are called into question. How do we create a legal duty of care for the earth? That was the big question that has driven my thinking. I looked at existing environmental and corporate laws and I saw they were not fulfilling this particular legal obligation. None of our existing laws set out a proper duty of care for the earth. We have a Universal Declaration of Human Rights, but the same does not exist for the earth. The earth has rights too, I reasoned, such as the right not to be polluted and the right to life. What if we had a similar Declaration for the earth, a declaration that gave formal recognition to the rights of non-human beings, such as the soil, the seas and the air we breathe. How much easier it would be for me, as a barrister, to represent my client the earth in court. Just as I can represent the unspoken words of a child because we impute human rights to them, so we can do the same for the earth.

EARTH RIGHTS

We may not have thought of other beings as having rights: however, they do exist. They may not be written down as formal laws in some jurisdictions, but to many natural law is a given. The right not be polluted is a right that belongs to the earth as much as it belongs to humans. To breach that right can be a result of neglect or an abuse. It can be an act or an omission; either by failing to do something, or by refraining from doing something, or by doing something that can result in damage, destruction or loss of ecosystems. Many of our existing laws are premised on permit allocation and limitations, not prohibition – these are laws that have proven themselves to be unfit for purpose.

Permits to pollute protect the polluter, not the earth. Fines levied after the event, when caught exceeding acceptable levels of destruction, can be sidestepped, litigated or paid-off. No amount of voluntary codes, environmental impact reports or energy efficiency targets will change matters until the concept of the 'environment as property', with ownership and thereby accrual of superior rights by the owner, is overturned. Slaves used to be property. It was argued that to present them with rights would be uneconomic, untenable, bring business to a halt. However, those businesses who profiteered out of slavery and sugar reinvented their wheels and not one went out of business as a direct result of the laws of abolition being put in place. This was in part because their slavery subsidies were replaced with subsidies which were for loss of business and to assist with facilitating new business that was not premised on the profiteering of slaves. Public pressure, mass petitions and recognition of rights for slaves combined to tip the balance and stop the trade. Laws were passed first in the UK, then other countries soon followed suit. Slaves were no longer another man's property, to use and abuse as he so wished.

No longer was it deemed acceptable to treat other persons as if they were items, to be bought and sold for profit. The shift that occurred when slavery was abolished was seismic; extrinsic values were replaced by intrinsic values. No longer was a human valued by his price tag; now a human was valued in and of himself. The ethical imperative trumped the economic imperative.

Slowing pollution levels by permit allocation just prolongs the inevitable problem; stopping the pollution at source changes business overnight. By making large-scale pollution a crime can stop further long-term damage from occurring. Prohibition is the inevitable next step; as has been demonstrated by current laws, small incremental steps are not going to get us there. We tried the small steps – now we need to take the leap. Pollution cases in the USA are being thwarted for lack of trans-boundary legislation. International law has a gap that needs to be filled.

In 2007 I researched the possibility of creating a new body of Earth Law. The outcome was an invitation from the United Nations to speak on my proposal for a new body of law, starting with a Universal Declaration of Planetary Rights. My proposal for a Declaration triggered a response that was the beginning of international engagement on the issue. I spoke at a UN Conference on Climate Change in November 2008, just after Ecuador had successfully voted by referendum to include in their new Constitution a Bill of Rights of Nature. The top twelve rights and freedoms were drawn up and were presented at a conference in Sweden the following year; just months later Bolivia decided to take it on and they opened up the process to the people. Thousands of people engaged in the process of drafting the Declaration and the outcome is the Universal Declaration of the Rights of Mother Earth which Bolivia is now taking to the United Nations. It will stand alongside the Universal Declaration of Human Rights and the new rights will create a legal framework from which other Earth Law can evolve.

Ecocide is the governing mechanism to protect the earth's right to life. By naming mass damage and destruction as ecocide, and by giving it legal definition, I realised we can halt escalating greenhouse gases at source, prevent further instability and prohibit dangerous industrial activity overnight.

APPOINTING GUARDIANS FOR DAMAGED LANDS

By way of analogy, in formative years a parent owes a duty of care to their child. His and/or her duty is to ensure the well-being of their child, for that child is utterly dependent on the parental care. As a mother or father, the duty as primary carer extends to others as more children arrive into the family circle. Motherhood and fatherhood are roles specific to ensuring the well-being of the child, a responsibility that diminishes as the child enters into adulthood. When a parent abuses their child, or fails to act to protect the child's interests, that parent has failed in their duty to their child. In recognition of the child's inability to defend him/herself, laws have been put in place to provide a remedy when a parent fails in their responsibilities. In such an instance, the court will appoint a guardian to represent the child, to speak on his/her behalf and ensure his/her well-being is addressed in the course of the proceedings.

Replace the child with the planet and the mother with a corporation – for instance a logging company in the Amazon – and a very similar scenario exists. The Amazon, like the child, is unable to speak of the damage that has occurred and the needs it requires to ensure future well-being. Unlike the child, it has no recognized rights in law and as a consequence no responsibility is identified as being owed by those logging the territory. If caught, the company will be fined for logging unlawfully, nothing more. Without the recognition of the Amazon's rights and the corporation's responsibilities to the Amazon, a guardian cannot speak on behalf of the territory in court and the individuals in the company cannot be effectively held to account. However, the well-being of humanity requires that those with superior responsibility in the company owe an over-riding duty of care to the territory within which they are working. Where that duty of care has been breached, the fiduciaries – the directors – have failed to fulfil their moral obligation to prevent unreasonable loss, damage and destruction.

In 1948 the United Nations created the crime of genocide in response to the mass atrocities which arose out of World War Two. Today we face mass destruction of the planet, but unlike genocide, ecocide is not a crime of intent. Ecocide is a crime of consequence, one that often arises out of the pursuit of profit without imposition of a legal duty of care. Currently there is no crime to address this anomaly during peace-time. War Crimes prohibit mass environmental damage, yet there is no law to stop the daily destruction that has become the norm for business. Corporate law dictates that profit determines activity, regardless of the consequence to others in the earth community.

GIVING A NAME TO THE PROBLEM

Ecocide can be the outcome of external factors, of a force majeure or an 'act of God' such as flooding or an earthquake. It can also be the result of human intervention. Economic activity, particularly when connected to natural resources, can be a driver of conflict. By its very nature, ecocide leads to resource depletion, and where there is escalation of resource depletion, war comes chasing close behind. The capacity of ecocide to be trans-boundary and multi-jurisdictional necessitates legislation of international scope. Where such destruction arises out of the actions of mankind, ecocide can be regarded as a crime against peace, against the peace of all those who reside therein – not just humans but of the wider earth community as a whole. In the event that ecocide is left to flourish, the 21st century will become a century of 'resource' wars.

During wartime environmental damage is already a crime. By extending the same provisions (the size exceeds 200 kilometres in length, impact on ecosystems exceeds three months, or severely impacts on human or natural resources) to ecocide, we can protect the earth from daily destruction in peacetime too.

There is an additional reason for seeking international recognition of ecocide: until we have identified the problem, we are unable to provide the correct solutions. International law evolves in response to the changing world, and is by no means a perfect beast, growing and changing direction as it expands. But

it is an arena that must develop, by necessity. Such is the extent
of ecosystem destruction with global consequences for us all that
principles and legal recognition on a par with genocide are now
urgently required for ecocide. Corporate-related destruction
and pollution clean-up determined by voluntary governance, we
know, has been manifestly unsuccessful. Creation of the crime of
ecocide creates a pre-emptive obligation to act responsibly before
damage or destruction of a given territory takes place. Thus, the
creation of the illegality in itself translates a moral obligation into
a legal duty. In doing so, the burden shifts dramatically, sending a
powerful global message to the world of a reinforced moral stance
for us all, not just in business, to take responsibility for the well-
being of all life.

I OWE VERSUS I OWN

When the old that does not work collapses, space is created to
make way for the new. It is at this juncture that we are afforded an
invaluable opportunity to put in place new systems that do work
and systems that provide resilience. However, as all architects
know, first we must ensure the correct foundations are in place.
The bedrock upon which we build is our choice: it can be solid,
one with the inherent values of the planet at its very centre. It
can be based on intrinsic humanitarian and ecological values,
values which belong to all of us and the planet globally. When
seeded from the outset, all else that comes thereafter will grow
and flow from them. These values will renew and shape the world
of our financial systems, our legal precepts, our governmental
bodies, the very decisions which we will make in our every day
life. These values and their corresponding rights can be placed at
the centre of all decision making, which will in turn shape our
behaviour and action for the future. Where our decisions are
not intrinsically-value based, then our decision-making will be
built on fast dissipating sands, which can do little to ensure we
create a better world. Climate negotiations demonstrate a non-
engagement on owing a duty of care, instead the focus was on
ownership. Markets to buy and trade trumped any dialogue about

owing a duty to ensure the climate is restored. The very fact that we have a 'negotiation' about a global issue, as if it were a market trading house where everyone has a different pricing scheme, is an indication of just how far removed we have become. Our climate is not for negotiating as a commodity; it is our duty to create a system that guarantees its and our health and well-being.

Where it is accepted that environmental imbalance (and the consequential escalation in greenhouse gases) is largely the result of human-created pollution, then it raises both a legal and moral argument that we have a duty of care to ensure that the pollution must stop. The introduction of contaminants – be they synthetic or an excess of natural – into the environment can lead to a point that causes instability, disorder, harm or discomfort to physical systems and living organisms: they are then termed 'pollutants'. For example, noise pollution – such as at a late night party in your neighbour's house will only be remedied when the music is turned off. Your sense of well-being is restored and a good nights sleep is no longer impeded. Likewise, polluting the atmosphere: stop the pollution at source and the issue is remedied. Reduce the pollution and you are the recipient of low-level but continual build of irritant; in the example of your neighbour playing music all night every night you may find the pressure begins to build. Likewise with pollution of the atmosphere – the pressure continues to build. Stop the ecocide at source, and offsetting, carbon crediting and carbon capture and storage becomes redundant. No longer is the pressure re-directed elsewhere, leaving a legacy for future generations to contend with. It is not the risk of the potential ecocide that is carried which needs to be evaluated, but the consequences when it goes wrong. Evaluation based on consequences will give a fuller understanding of the outcome we leave for others to resolve. To leave a problem unresolved for others to work out is to leave our world in a worse place than when we arrived. Alternatively we can choose another route: one that leaves our world with a legacy of life and systems that do not store problems for the future.

Justice encompasses ecological justice (as argued by jurists for centuries). If we accept that premise, then peace must also apply to the functioning of all those that reside within a territory – both people and planet. To ensure ecological justice is to ensure that ecosystems remain intact and functioning. When ecosystems malfunction, we have system breakdown and ecocide which in turn leads to resource depletion, and then to conflict and ultimately war.

ABDICATION OF RESPONSIBILITY

Three days before the conclusion of the 2010 UN climate negotiations Ban Ki-moon spoke at a private audience in the luxurious confines of the beach Hotel Marriott. Sharing the platform was the President of Walmart, the largest corporation in the world, who received rapturous applause when he declared himself an environmentalist. 'Every second of every minute of every day a football pitch-size of forest is destroyed' we were informed. Yet despite seeking reliance on this sobering fact, destruction of the earth was not recognized to be a crime. Ban Ki-moon, the head of the UN and the man with ultimate superior responsibility to ensure that a solution was found by the conclusion of the climate negotiations, rushed off to fly out of Cancun that afternoon safe in the knowledge that the negotiations were to prove futile in the face of such enormous mass ecocide. Ban Ki-moon's departure was symbolic of the lack of engagement by those in a position of superior responsibility. Instead, the planning officials acting on behalf of the people and our planet – the governments and their negotiators – were advancing the interests of big business by the creation of mechanisms and trade deals to secure rights over the worlds forests. There is a 'growing acceptance', as *The Economist* commented, 'that the effort to avert serious climate change has run out of steam.'[*]

---

[*] 'How to live with climate change', *The Economist*, 25.11.2010

BREACH OF OUR HUMAN RIGHT TO LIFE

Confining ourselves to breaches of human rights fails to take in the wider picture. Climate change cases are failing at the court door for being argued on Human Right breaches. For the crime of ecocide however, the damage is to the environment and all who live there (or who are impacted by the ecocide). Human damage can be and often is secondary (it arises out of the primary damage – for instance the polluted waters can give rise to illness). Thus the *actus reus* (the physical element – the doing of the act) of the crime is the establishment of damage to the *territory*, not specifically the doing damage directly to the people. For instance, it is not the people who are destroyed, but the land on which they live and, as a result, their health and well-being is compromised. There is a crucial point here that has until now been overlooked: our human right to life is placed at risk when damage, destruction or loss of ecosystems occurs. It may be that the risk to life is not immediately apparent, in so far as people are not exhibiting injury or pain, however the injury to the persons can manifest later or in a people in another territory.

Thus, a belching power-plant may cause injury to as yet unborn children. Furthermore it contributes it's pollution to the atmosphere and along with other industrial activity builds a dangerous legacy for persons unknown in other territories. Ecocide can be a crime against the right to life of not only current beings, but also unknown and future generations.

It is a death by a thousand cuts. Each time a cut is accepted, our lives are compromised bit by bit. No one cut can be said to be the final determinant, as a whole the overall impact can be viewed as a risk of such enormous magnitude to human life that we have little choice but to take the route to outlaw it. We haven't done this yet, but the arguments to do so are already in place. Morally we have no choice if we are to uphold the right to life for future generations.

Human Rights case law demonstrates that our EU states are in breach of their duties to EU citizens where an EU State fails to protect the human right to life. Failing to protect our right to life includes failure to prevent injury and risk to health which arises from dangerous industrial activity.

In 2009 one crucial term was agreed at the climate negotiations, held in Copenhagen. To prevent dangerous anthropogenic interference with the climate system, the Copenhagen Accord[1] recognized 'the scientific view that the increase in global temperature should be below 2 degrees centigrade', in a context of sustainable development, to combat climate change. In other words, 'dangerous' was accepted to mean a 2 degree (or more) increase in temperature.

Under Article 2 of the European Convention on Human Rights (ECHR), our Right to Life is defined as the following:

'Everyone's right to life shall be protected by law. No one shall be deprived of his life intentionally save in the execution of a sentence of a court following his conviction of a crime for which this penalty is provided by law.'

Our right to life is a statement of principle. Statements of principle require interpretation by courts to bring meaning when applied to the facts of a given situation. Current ECHR case law governing our right to life under Article 2 has this to say: 'any activity, public or private in which the right to life is put at risk, especially dangerous industrial activities, must be prohibited where there is knowledge that the risk to life and the duty to stop is known. Authorities which knew or ought to have known of the risks are, where they should have taken measures and did not, in breach of Article 2.[2]

In other words, governments and corporations who know that their activities are contributing to placing humanity at risk of danger to life must halt their dangerous activities. Failure to do so places them at risk of prosecution under Article 2 of the ECHR. Placed within the context of climate change, this case law is a ruling which could be found to be binding on all EU nations. It is a case that could change the rules of the game.

The implications for companies that own power stations, generate or sell energy from fossil fuel extraction, are huge. All

---

1 The Copenhagen Accord is a document that delegates at the 15th session of the Conference of Parties (COP 15) to the United Nations Framework Convention on Climate Change agreed to "take note of" at the final plenary on 18 December 2009.
2 *Oneryildiz v Turkey* [2004] ECHR 657

energy companies contribute enormously to the generation of excess greenhouse gases which in turn exacerbate climate change. Governments already accept that excess greenhouse gases are caused by human activity, it's just a matter of examining which activities are the most damaging of all. The major contributors are those in the extractive industry who generate a substantial%age of the overall excess. Contribution to the pool of greenhouse gases by any industrial activity that places human life at risk can be said to be ecocide.

Dangerous industrial activity is classed as a criminal act and as an offense against persons, according to ECHR case law. Where this is the case, authorities have a legal duty of care to take preventive operational measures to protect individuals whose lives are at risk from the criminal acts of another individual.[3] States, EU case law tells us, must put in place legal and administrative mechanisms to deter the commission of offenses against the person.[4] Under existing case law, the time is ripe for new laws to be put in place.

Law shapes our societies, our way of thinking, our behaviour. By labelling our world a thing of property, legal systems have legitimised and encouraged the abuse of Earth by humans. Now we know the problem, we have a responsibility to put it right.

THE END OF THE ROAD

Climate negotiations have reached a critical point. On the current trajectory, scientists now estimate that a four to seven degree centigrade increase in temperature is now certain. The Kyoto Protocol is unlikely to mitigate to any great degree the onslaught of climate refugees fleeing the land most likely to be destroyed or lost to catastrophic rising sea levels, floods and storms. None of this will be remedied by a trade in carbon. Neither will it prevent the loss of biodiversity and species extinction that is sweeping across all nations with the speed of a forest fire. Grassroots organizations and citizens from all over the world mobilized in Cancun to call for putting people and planet first. They had marched and protested peacefully

3   *Osman v UK* [1998] 29 EHRR 245
4   *Makaratzis v Greece* [2004] 41 EHRR 1092

but their voice was not heard. The day before the conclusion of the summit a stadium in Cancun was filled with 5,000 people who had travelled from all corners of the globe to voice their concerns at the negotiations. Heavily armed militia had ensured that civil society stayed out of sight of those they wanted to communicate with and as one negotiator confided to me, their protests were either unknown of or simply ignored. Only one leader went to hear what the people had to say and to speak with them: that was Evo Morales.

Evo Morales is a leader of 10 million people in Bolivia where thirty-seven indigenous languages are spoken. He is himself the first indigenous leader to be elected president in South America. As an Aymara Indian president he is changing the politics of climate change. Morales is also titular president of Bolivia's *cocalero* movement, a federation of coca growers' unions, who are resisting the efforts of the United States government to eradicate coca in Bolivia under the United Nation's (UN) 1961 Vienna Convention on Narcotic Drugs. The coca leaf is part of everyday life for people in the Andean region. An estimated seven million people in a region stretching from southern Colombia through Ecuador, Peru, Bolivia, northern Chile and Argentina chew coca leaves, as did their ancestors going back many generations. Renowned for its stimulating and blood oxygenation benefits, each leaf is overflowing with vitamins and numerous health-giving alkaloids. Regardless, the Narcotics Convention designates the coca leaf as a narcotic, and has banned the growth and use of coca. Coca does indeed release a mild narcotic, which serves to combat altitude sickness, hunger and fatigue when chewed. Its use in a country where altitude sickness, hunger and fatigue are every-day problems is essential for social cohesion and functioning. Bolivia's new Constitution describes coca as a 'cultural heritage, a renewable natural resource' and a key biodiversity element that helps maintain Bolivian well-being. Nevertheless, the United States continued to oppose the Bolivian proposal to lift the ban on using coca for medicinal purposes. Morales has now withdrawn his country from the Vienna Convention on Narcotic Drugs. His decision was based on the fact that the Convention contradicted Bolivia's 2009 Constitution. This bold move puts indigenous rights in the limelight and underlines the anachronistic and discriminatory nature of the 1961 Convention, as well as the need to revisit this treaty in order to create a more appropriate international law directed towards coca chewing. Whether these actions will alienate Bolivia in the international arena remains to be seen, but Bolivia's bold withdrawal from the Convention actually may prove beneficial in raising awareness both regarding coca chewing and indigenous rights. This single act highlights the injustice of a system that is insensitive to cultural differences. The Convention is flawed and fails to protect fundamental indigenous rights: as such it is no longer fit for purpose.

CLOSING THE DOOR TO DANGEROUS INDUSTRIAL ACTIVITY

On the 15 of March 2012 the Organisation for Economic Cooperation and Development issued a stark warning; carbon dioxide emissions from energy use are expected to grow by 70% in the next thirty-eight years because of our dependence on fossil fuels. As a result, by 2100 the global average temperature will have increased to between three and six degrees centigrade.[5] The causal link has been established, the evidence is there for all to read; conventional energy production and use will take us over the danger line. The continued use of fossil fuel is dangerous and risk of injury to human and non-human life is real and immediate. In Europe we already have a legal duty of care to close down any dangerous industrial activity and morally all nations have the same duty too. Humanity is faced with a choice: continue with 'business as usual' or confront the urgent need to adapt.

Our existing global policies have proven no longer to be fit for purpose. Emergency measures are called for, to create a global stabilisation policy. The Law of Ecocide provides a framework for intervention to stop dangerous industrial activity that causes increase in carbon dioxide emissions, to disrupt 'business as usual' and to act as a bridge to the green economy.

---

5   http://www.oecd.org/document/11/0,3746,en_2649_37465_49036555_1_1_1_37465,00.html

Chapter 2

THE EMPEROR HAS NO CLOTHES ON

CANCUN, Mexico, 30 November 2010, the 16th round of the United Nations Framework Convention on Climate Change (COP 16) has commenced. Negotiators, lobbyists, scientists, NGOs and a whole host of non-accredited persons have descended upon the Mayan beach resort. There was much to disagree about, but surprisingly even more to agree upon. The common ground that all parties held was the acceptance that climate change is exacerbated by human activity and the urgent need to curb activities which are damaging and destructive. In this respect, most politicians, environmental and social organisations have much in common. If they were to gather their strengths, their collective input could be a true force for change. But with a focus that remained solely geared to their own respective agendas they all failed to address the single most important issue of all.

Instead politicians, environmental and social organisations disagreed and clashed repeatedly. They could not even unite over the date for a demonstration. The year before at COP 15 in Copenhagen, 100,000 people had walked peacefully from the city centre to the conference hall where the negotiations were being held. In Cancun two groups took differing routes outside on the streets whilst the NGOs, lobbyists and negotiators stayed well away, believing it safer to remain inside and enjoy their regulated, air conditioned sanctuaries.

The climate negotiations use science as their primary source of evidence to establish the parameters of danger. An extensively laborious and time consuming system of engagement is used whereby over 2,000 of the world's top scientists come together every five years to agree a range of scenarios, based on the most up-to-date scientific reports. They work from an agreed set of factors that are based on non-disputed presumptions. All governments take this as their starting point to evaluate the implications of damage and destruction that will occur. It is for them to evaluate what has to be done to mitigate the extent of disruption to society. Therefore it would seem logical that both environmental and social organisations are required to work together with the governments and join forces in addressing the implementation of plans that are required. It should be the annual meeting-place for all three stakeholders; political, social and environmental, to meet and build the basis of their collaborations which can then be continued in their countries thereafter. Only through collaboration and the passing of laws, such as the criminalisation of ecocide, can newly established systems be embedded to ensure successful transition for all communities.

The United Nations was originally set up as the mouthpiece for the Peoples, not governments, nor business. Somewhere along the line its role has been reversed, and the Peoples have been left firmly outside in the sweltering heat of every-day reality of living with climate change, whilst those who are representing the People sit coolly discussing matters with heads of business to determine how they can profiteer from the situation. Corporations have access to the politicians and Heads of State where the man on the street does not. As a result, climate negotiations have become mired in a tussle over who retains most power whilst the ship sinks. The irony is that the People are ignored and forgotten about. The UN has a duty to the People, but that duty has been ditched in favour of business interests and political leverage. Harking back to the true origins of the UN, a proper process of addressing climate change would be to hold a Peoples' Summit where the host country provides an open space in which civil society can follow and be a participant in the proceedings. That is true democracy. Democracy cannot survive without knowledge and there can be no doubt that solving the climate crisis involves engaging the People at every level.

Mexican Foreign Minister, Patricia Espinosa, the Summit President of the United Nations Climate Change Conference (COP 16) announced the successful conclusion of the two weeks of wrangling. The final text of the United Nations Programme on Reducing Emissions from Deforestation and Forest Degradation in Developing Countries (REDD+) had been agreed. Facing a hall full of negotiators and representatives from 194 countries, she announced that the deal had been sealed: REDD+ had been orally signed off by all 194 Parties, bar one. Her words were met with cheers, whoops and a standing ovation with people clapping long and hard. Negotiators, leaders and their assistants were smiling, hugging and shaking hands. The desire for a positive outcome that signalled success, no matter how small, was unmistakable: the sense of elation was palpable and infectious. Only one man thought otherwise. He stood separate from the party spirit and looked on with disbelief. He had described the conclusion as 'Governmental ecocide.' Evo Morales is the man who spoke out: he is the president of the Plurinational State of Bolivia. He had refused to accept the outcome but his protests had been over-ruled. Evo Morales is often derided for wearing the clothes of his people; sometimes that includes a poncho and trainers rather than a suit more commonly preferred by Western leaders. He was the only leader to publicly speak out against the so-called solution. No-one else was prepared to stand at his side and support his words, not even his fellow Latin American leaders. What he had to say was not what those in the hall wanted to hear.

The core fundamentals of REDD+ are set out in a document drafted by the Ad Hoc Working Group on Long-term Cooperative Action under the Convention (AWG/LCA). Developing country Parties can contribute to mitigation actions by trading their forests in whichever way they deem appropriate within the strictures set out under the following five headings:

(a) Reducing emissions from deforestation;
(b) Reducing emissions from forest degradation;
(c) Conservation of forest carbon stocks;
(d) Sustainable management of forest;
(e) Enhancement of forest carbon stocks.

In truth, the remit of conservation has been reduced to a single-issue: carbon trading. A market has been built on the transference of ownership of the carbon to be found in forests. Forests are now viewed simply as 'carbon stocks' rather than ecosystems. Buried in Annex 1 is a non-binding consideration for the people who live and depend on the forests under the agreement. It is designed to quell the concern of those who insist on their rights being protected. The thorny issue of indigenous rights can be ignored by those trading the forests, unless they feel inclined to do otherwise. Such fettering of the system has been relegated to a voluntary decision: each trader can if he so wishes, promote and support the rights of those who live and survive within their forests. The REDD+ document 'requests developing country Parties' to ensure the 'full and effective participation of relevant stakeholders, inter alia indigenous peoples and local communities'. Sounds official, but read carefully: the wording is carefully chosen to ensure that such participation is non-binding. There is no provision to monitor whether any participation is applied, and when applied whether the stakeholders were given more than a token role. Moreover, the definition of forests is wide-ranging. It includes decimated land, industrial mono-crop plantations and genetically engineered trees, all which can be traded for carbon permits. In essence, if you have a tract of land which may or may not have trees growing on it, the air above it can be traded providing you hand over the rights to that land to the traders. The ongoing destruction, pollution and loss of ecosystems does not stop. Continued abuse comes at a small price to the polluter but at a price that has no limit to the rest of the world.

THERE ARE THREE COMPELLING REASONS WHY GOVERNMENTS OUGHT TO IMPLEMENT A LAW OF ECOCIDE:

1. to create the legislative framework for a green economy
2. to create jobs and build resilient economies
3. to gain electoral support.

During war-time, it is already an international crime to destroy the environment. During peace-time no such crime exists. By making a law that effectively ends extensive damage and destruction to the environment during peacetime, human-made ecocide can be outlawed for all time. Most corporate ecocides are crimes of consequence – the damage is something that few consider as an issue to prevent and, if considered, is treated as an externality to be factored in as an external cost. The damage is rarely intended, it is considered a by-product of decisions that maximise profit without looking to the consequences. Making ecocide a crime without intent will prevent years of unnecessary litigation and costs.

Ecocide is a law to stem the flow of destruction from the outset. It is an upstream solution, far more cost-effective than costly downstream overheads after the damage has been done. Moreover, it is a law which will create a level playing field across the world.

We now have the knowledge that fossil fuel is no longer serving humanity well. To continue it's use is tantamount to a crime – of ecocide. To do so will put humanity at jeopardy; and not just humanity, but the very life of Earth itself. Catastrophic climate change is a risk we have to face; no longer is it safe to use fossil fuel. It is crucial to change our course of action. Once we accept that we can no longer continue business as usual, we can create the legislative framework to ensure a rapid and smooth transition. What is needed is adaptive leadership; leadership that can adapt to the size and magnitude of change envisaged. Leadership that is bold, moral and courageous. Not one Member State can justify putting humanity at risk when the whole of civilisation stands on the brink of disaster.

Carbon emissions are just one of the adverse impacts of dangerous industrial activity. There are many more; soil erosion, pollution and biodiversity are all on the brink of triggering mass crises. International law, laws that are superior to national laws, can impose a new system which changes the rules for us all. A Law of Ecocide will close the door in one direction: only when we do that can we open another door in a completely different direction. It will take courage. It will take adaptive leadership.*

---

* Higgins, P., Closing the door to dangerous industrial activity: a concept paper for governments to implement emergency measures, 21.03.12.

....

WHY WE NEED A LAW ON ECOCIDE[1]

Sixty-six years ago Sophie Scholl and nineteen others were sentenced for their activities and words against the tyranny of the Nazis; on 5 January 2011 Dan Glass and nineteen other climate campaigners were sentenced for their activities and words against the tyranny of E.ON. Scholl and her friends stood up and took direct non-violent action against what they saw to be a threat to humanity. Their crime was the dissemination of leaflets highlighting and decrying the tyranny of the Nazi dictatorship. It was a decision to undertake something unlawful – an act that they believed was a necessity – to halt a greater but unnamed crime, a crime that cost many lives. That crime did not, at the time, have a name. But it soon did: genocide.

Dan Glass and his friends did the same in April 2009. They too were prepared to stand up and take action. Their crime was planning to shut down Ratcliffe-on-Soar, a coal-powered station owned by E.ON, one of Britain's largest greenhouse gas emitters. The state was failing to prevent a greater injury from taking place; the loss of life. This time it is not only human life, but all life.

Like Scholl and her friends, Dan and his fellow defendants were motivated to take non-violent direct action. It was a decision to undertake something unlawful – an act that they believed was a necessity – to halt a greater but unnamed crime, a crime that will cost many lives. That crime does not yet exist. But it does have a name: ecocide.

---

1   Higgins, P., 'Why we need a law of ecocide', *The Guardian*, 05.01.11

Currently there is no law to prosecute those who are destroying the planet. Instead, climate campaigners do not have the support of the judiciary in preventing the corporate ecocide that is daily occurring under our very noses. Corporate ecocide has taken us to the brink of collapse: excess greenhouse gases from heavy industrial activity place the lives of many millions at risk from human-aggravated cataclysmic tragedies.

Over the passage of time, tyranny revisits. Tyranny is the cruel, unacceptable, or arbitrary use of power that is oblivious to consequence. While the use of coal stations may not be deemed an intentional cruelty, it is certainly an unacceptable use of corporate power. Our governments collude by encouraging excess emissions, contrary to their UN commitment to stabilise 'greenhouse gas concentrations in the atmosphere at a level that would prevent dangerous anthropogenic interference with the climate system'.

Sixty years ago the tyranny was Nazism. Today it is pursuit of profit without moral compass or responsibility. Despite the planned Ratcliffe protests, dangerous industrial activity continues and is actively encouraged while the majority of humanity accepts it regardless of the known consequences.

The failure rests with us all. Our governments are unwilling to intervene to make the destruction of our world a crime; our police do not have the laws to prevent the ecocide and our justice system is unable to protect our greater interests when faced with the superior silent right of corporations to cause injury to persons and planet. Those who stand up and speak out are thereby treated as criminals.

In Britain, our citizens have the right to argue their case. In Nazi Germany, the individual, as well as the state, was viewed as subservient to the race and therefore had no say. The language simply didn't exist to give name to the crime they were unable to speak up against – or if they did they were met with a punishment so excessive that virtually all were silenced through fear and guilt. The primary purpose of Hitler's Naziism was to uphold the Aryan race as supreme above all else. In Germany, the Nazi movement had come to power through the pursuit of national unity by the

forced repression of national enemies and the incorporation of
all classes and both genders of Aryan descent into a permanently
mobilized nation. A race, Hitler believed, that would strive to
annihilate all enemies by all ways and all means.

Unlike Sophie and her friends, Dan and his friends were
granted their right to argue their case. Expert evidence was heard,
from James Hansen, the former head of NASA's Goddard Institute,
on the immediacy of the threat to life caused by the escalation of
emissions, and fromMPs who confirmed government inertia. All
of which the jury failed to accept or that there was a greater risk
at stake – one that posed a threat to all of humanity. Unlike the
Ratcliffe twenty, Scholl and her co-conspirators were denied the
right to defend themselves in their trial. They too were convicted
for resorting to unlawful acts, which they believed to be necessary
to expose the truth. At the very end of her trial, Sophie spoke out.
It is just matter of time, she said, before the true destroyers are put
in the dock. This was only a few years before charges of genocide
were brought at the Nuremberg War Trials.

National Environmental Legacy Act (NELA)

A Climate Legacy Initiative[2] Policy Paper on intergenerational
ecological justice and share mechanisms recommends that we
arm national and international legal systems with laws to respond
to the threats of climate change. The authors describe the existing
United States legal system as a product of an industrial society, a
society that was rich with abundant natural resources and that had
few laws to curtail their use. A new legal framework is necessary,
argue its authors, for addressing the adverse impacts of climate
change and for protecting the interests of future generations where
natural resources have become scarce or are at risk of becoming
extinct. Access to and development of natural resources for all
future generations addresses intergenerational justice concerns
such as the ability of generations not yet born to have enforceable
legal rights.

---

2   see http://www.vermontlaw.edu/Academics/Environmental_Law_Center/
    Institutes_and_Initiatives/Climate_Legacy_Initiative/Publications.htm for full text

One proposal is for a National Environmental Legacy Act (NELA). A NELA would codify public natural resources as a legacy duty for future generations. The federal government would identify the legacy it wishes to leave to future generations and provide rules for achieving that goal. The authors' recommendations include prohibitions on impermissible levels of environmental degradation or depletion, as well as provisions for collection and measurement of scientific data, regulation implementation and enforcement.

INTERNATIONAL COURT OF THE ENVIRONMENT (ICE)

One of the contributors to the 1992 Rio Earth Summit was The International Court of the Environment Foundation (ICEF). They proposed the creation of an International Court of the Environment at a global level. Also on 9 March 1992 the European Parliament, with the impending Rio Earth Summit in mind, formally adopted a Resolution on the constitution of an International Tribunal for the Environment. However, the European Parliament Resolution was not made public at the first Rio Earth Summit.

More recently, the idea has gained much ground; the Italian Ministry of Foreign Affairs hosted a Global Environmental Governance Conference on 21 May 2010 pulling together many of the parties who have over the past twenty years been working on the proposal. ICEF advocate the promotion of a balanced system of global environmental governance through the creation of an International Environment Agency to monitor and manage the environment and the formation of an International Court of the Environment. By giving member states, associations and individuals access to a new International Court of the Environment (ICE), or through the extended jurisdiction of the already constituted International Criminal Court (ICC), they hope to secure the conviction of international environmental crimes. Creating a court of environmental law is one route to ensuring effective governance, monitoring and management of the environment.

The International Criminal Court (ICC)

In 1998 the Rome Statute as adopted. It set up the International Criminal Court to investigate and examine the four Crimes Against Peace when States fail to take action. It is an independent institution financed by the State Parties. It is not part of the United Nations and State Parties cannot interfere in the judicial functions of the Court. The Court is governed by an Assembly of States Parties and is composed of four organs: the Presidency, the Judicial Divisions, the Office of the Prosecutor and the Registry.[3]

Crimes Against the Earth are not yet included within their remit, save for environmental War Crimes (Article 8(2)(b)). Ecocide has yet to be included as the 5th Crime Against Peace, and the Earth's Right to Life has not yet been added. It is a matter of time before these laws are included within the remit of either the ICC or its off-shoot, a Court for the Environment. An International Court for the Environment (ICE) has been the subject of enquiry for a number of years and has been examined in detail by many in the legal profession. What is clear is that we need a specialized Court with jurisdiction over individuals as well as States and transnational corporations which have so far been able to escape public scrutiny. Whether it is created as an extension of the existing ICC or set up on ICC rules and procedures, matters not – what is crucial is the house within which to place the new laws.

The ICC has 119 Member State Parties. However, a number of states, most notably China and India are not signatories[4]. The ICC is the first permanent international criminal court established to prosecute individuals for the most serious crimes such as genocide, crimes against humanity and war crimes. The Court is governed by the Rome Statute which came into force on 1 July 2002 and can only prosecute crimes committed on or after that date. The official seat of the Court is at The Hague, Netherlands.

The ICC has universal jurisdiction. It steps in when national courts are either unwilling or unable to investigate or prosecute

---

3   Detailed structure can be found at http://www.icc-cpi.int/Menus/ICC/Structure+of+the+Court
4   As of 08.11. For updated list of State Parties see http://www.icc-cpi.int/Menus/ASP/states+parties

nationals for Crimes Against Peace. Since 2006, the ICC has opened investigations into six cases of international crime in the Democratic Republic of the Congo, Uganda, the Central African Republic, Sudan, Kenya and Libya[5]. Of these six, three were referred to the Court by the State Parties (Uganda, Democratic Republic of the Congo, the Central African Republic); two were referred by the United Nations Security Council (Sudan and Libya); the sixth (Kenya) was begun *proprio motu* (of his own motion) by the Prosecutor. Additionally, the Prosecutor has requested a Pre-Trial Chamber to authorize him to open another *proprio motu* investigation.

The Court has as of August 2011 publicly indicted twenty-six people, proceedings against twenty-four of whom are ongoing. The ICC has issued arrest warrants for seventeen individuals and summonses to nine others. Five individuals are in custody and are being tried while eleven individuals remain at large as fugitives. Proceedings against two individuals have been closed following the death of one and the dismissal of charges against the other. To date, the Office of the Prosecutor has received over 9,000 communications about alleged crimes. Around half of these communications are dismissed as 'manifestly outside the jurisdiction of the Court'.[6]

Under the Rome Statute's complementarity principle, the court is designed to complement existing national judicial systems: it can exercise its jurisdiction only when national courts are unwilling or unable to investigate or prosecute such crimes. Primary responsibility to investigate and punish crimes is therefore left to individual states. Thus many states (but not all) have implemented national legislation to provide for the investigation and prosecution of crimes that fall under the jurisdiction of the ICC. Hence, by implementing ecocide as a crime at international level, the capacity exists for the crime to filter rapidly into national legislation and consciousness.

---

5 As of 08.11. For updated list see http://www.icc-cpi.int/Menus/ICC/Situations+and+Cases/ Situations

6 As of 09.10 http://www.icc-cpi.int/Menus/ICC/Structure+of+the+Court/ Office+of+the+Prosecutor/Comm+and+Ref

THE EIGHT POINTS OF DEEP ECOLOGY BY ARNE NAESS

1. The flourishing of human and non-human life on Earth has inherent value. The value of nonhuman life-forms is independent of the usefulness of the nonhuman world for human purposes.

2. Richness and diversity of life-forms are also values in themselves and contribute to the flourishing of human and non-human life on Earth.

3. Humans have no right to reduce this richness and diversity except to satisfy vital needs.

4. The flourishing of human life and cultures is compatible with a substantial decrease of the human population. The flourishing of nonhuman life requires such a decrease.

5. Present human interference with the nonhuman world is excessive and the situation is rapidly worsening.

6. In view of the foregoing points, policies must be changed. The changes in policies affect basic economic, technological and ideological structures. The resulting state of affairs will be deeply different from the present and make possible a more joyful experience of the connectedness of all things.

7. The ideological change is mainly that of appreciating life quality (dwelling in situations of inherent value) rather than adhering to an increasingly higher standard of living. There will be a profound awareness of the difference between big and great.

8. Those who subscribe to the foregoing points have an obligation directly or indirectly to participate in the attempt to implement the necessary changes.*

---

* Naess, A., *The ecology of wisdom: writings by Arne Naess*, pp.105 - 119. Norwegian Professor of Philosophy who lived by his philosophy to 'think like a mountain.' He gave name to the philosophy of Deep Ecology, the belief that our concern is for all of nature equally, including humanity.

RIGHTS AND FREEDOMS

The rights of non-human species are constantly evolving and gaining momentum (most recently Spain legally recognized rights of chimpanzees). In the UK we have a whole raft of regulatory duty of care laws for animals; whilst their rights are not always directly specified, they are implied. On an international level, the UN Convention on the Laws of the Seas implies rights for the seas – the right not to be polluted being the primary one.

The difference between rights and freedoms is rarely distinguished in law, but there is a fundamental difference. A right is a moral or legal entitlement to have or obtain something or act in a certain way, whereas a freedom is the state of not being subjected to or affected by, without restraint. Enshrined in international law we have certain human rights and freedoms, such as the right to life, the freedom from slavery. However, the human right to a healthy environment does not exist as a human right under the Universal Declaration of Human Rights, neither do we recognize in our international laws the rights of nature. Such rights do exist, they're just not written down in most nations' constitutions or legislation. Rights such as nature's right to restorative justice are beginning to be recognized in some jurisdictions, as well as the human freedom of a healthy environment. Ecuador has enshrined in their Constitution the Rights of Nature and Bolivia passed a Law of Mother Earth in 2010. The Philippines Constitution has enshrined the duties of the State to uphold their citizens' rights to well-being, to health and to a balanced and healthful ecology, and Norway's Constitution reads at s.110b: 'Everyone has a right to an environment that ensures health and a natural world/ nature/environment whose production capacity and diversity is maintained'. Other rights have been ascribed to humans but have been ignored or marginalized by those who have vested interests in the land upon which those people with rights live, such as the rights of indigenous people.

Self-determination is one principle enshrined in law and embedded in the Charter of the United Nations, the founding document of international law. Self-determination applies to people who are of a given grouping or ethnicity that implies social cohesion and unity. It can also apply to a community of individuals who have come from a specific geographic area that they have identified as the boundary of their territory. These rights should in law protect the people and the territory, for example when future decisions are made by others who have a financial stake in the territory. State sovereignty is frequently invoked as justification for refuting the rights of indigenous peoples to self-determine. Now, corporate sovereignty can be added to the list of arguments used against the human right of indigenous communities to self-determine their outcomes and the outcome of their habitat. In reality, the attempts to create a carbon market have caused a form of cultural genocide. Those who are proving to be uncooperative, usually the indigenous community who live on the land, can be simply ignored.

THE FREEDOM OF A HEALTHY ENVIRONMENT
One particular freedom that was not included the Universal Declaration of Human Rights was the (human) freedom of a healthy environment. More specifically, such a freedom relies on two rights being adhered to: firstly, the right not to be polluted and secondly, the right to restorative justice. When both are applied, one can enjoy the freedom of a healthy environment. An increasing number of lawyers now believe that such rights should apply not only to people, but also apply equally to the natural environment as a whole. Indeed, there are a raft of other rights that have not been elucidated in our international legislation that apply equally to all beings, not just humans.

By creating laws that govern our relationship with nature, we can facilitate the creation of a framework which firmly establishes an ethical relationship between humans and nature. This is the beginning of thinking in less homocentric terms. In so doing we protect the natural world not only for ourselves but also for future generations. Whilst we know that the environment does not exist solely for man, it is possible that man exists for the environment. The right to life is, in truth, independent both of international law and of the municipal laws of States. It pre-exists law. Mohammed Bedjaoui, the former President of the International Court of Justice has this to say:

In this sense it is a 'primary' or 'first' law, that is to say a law commanding all the others. In this field, the simplest ideas are probably the most relevant. It would not occur to any lawyer, whether positivist or otherwise, to maintain that international law contains principles that are contrary to the elementary right to life. Thus the right to development imposes itself with the force of a self-evident principle and its natural foundation is as a corollary of the right to life. If the right to development does not, on this basis, belong to universal compelling law, it would have to be concluded, by the same process of reasoning, that genocide, which is the negation of the right of peoples to life, is permitted by international law.[7]

In essence, Bedjaoui is positioning the right to life as central to all other considerations and all other laws flow from there. The right to development is a right that whilst not explicitly stated in the Universal Declaration of Human Rights, can be argued to be an offshoot of the right to life where development does not constrain the life of humans. Bedjaoui makes the point that genocide, being a negation of the right to human life, is a basic tenet of international and as a consequence is superior law.

The right to life is our fundamental human right, from which all other rights and freedoms stem. Take our lives out of the equation and all else in our man-made world topples. The same applies to our land, our world. Take the life of planet Earth out of the equation and all else in our natural world topples. Without the life of our planet our human made constructs count for nothing. Preservation of our world should therefore be our number one concern, if we are to successfully navigate the waters of our very existence. Without life on Earth, we have no means of survival. Our very life-force is taken and replaced with inert matter that cannot replace that which is no longer alive. Species extinction cannot be simply replaced by machinery or by genetic manipulation. Some losses are now inevitable, however, we can prevent further losses. We can halt the march of destruction by taking decisive steps and

---

7   Smiley, D. *Crumbling Foundations, How Faulty Institutions Create World Poverty*, 2010, p.214

closing the doors to all that negates the life of Earth. That means destructive business practices will have to stop. To make a profit from systems that continue to destroy the earth and ultimately us no longer makes good business sense. There are other ways and means of creating wealth and jobs which can guarantee the life of our future generations. We have many business opportunities to create the new world we want to inhabit.

There is no doubt that this will entail a turnaround on a scale the likes of which has never been done before. But we have much to help us in this regard. Businesses come and go with increasing regularity – to stay ahead of the game requires an understanding of where the world is going. The law is an ever-evolving field of governance. It echoes our progress as a civilization. The true hallmark of an evolved civilization is one which takes the leap before the push comes. We are about to be pushed. Let's take the leap.

Chapter 3

NATURE AS A COMMODITY

POLLUTION travels. Pollution is understood in law to be the introduction of contaminants into an environment that causes instability, disorder, harm or discomfort to the ecosystem. The instability, disorder or harm can occur many miles away from the source of the contaminant being released. For instance, DDT released in the 1970s (and now banned) is turning up in the breast milk of inuit mothers and marine life throughout the world, many years after the ban and use of it in the USA and Europe. Mercury is another example: mercury has the ability to travel indiscriminately. Melting ice in the Arctic has been shown to release mercury thought to have been stored for decades. As the ice melts it is re-released back into the food-chain, storing in the fat of polar bears and other animals. Logically mercury should not reside there. With every cremation of a human with mercury fillings that have not been removed, mercury enters into the atmosphere and travels far.

PROPERTY LAW IS KING

Creating property out of the earth's resources – that which has been here for far longer that we have – has practical and long-term consequences for the future of Earth as a living being. Where we create property rights we create ownership: 'We *own*' as opposed to 'We *owe* a duty of care'. This is the difference between ownership and stewardship. Stewardship is governed in law by trusteeship laws. Ownership is governed in law by property law. Property is the ownership of any entity by a person or group of people who buy the rights to rent, sell, exchange or consume. They are perfectly within their rights, as owners to destroy and/or prevent others from intervening on the behalf of that which they own.

In other words, in law, the owner of the property can enforce his or her own rights over and above any perceived rights that may pre-exist. Rights of ownership give power to the owner, without consideration of the rights of various other parties, including that which they have just bought. Property is not a relationship between an individual and a thing, but in fact it is a relationship created in law between people and people. A document created by lawmakers, or governments, which sets out the contract of ownership is all that is required to sanction the transfer of ownership into the hands of another.

Changes are tangible when systems of private property are introduced, for instance when land is divided into packages to be sold to the highest bidder, or that particular package of the earth's carbon-cycling capacity is accorded a price. Those that gain from the system are those who can afford to buy-in to the system. The newly created owner, group or company, who have bought the rights over the land are the ones holding the new privileges. A good example of how this system has effected dramatic change to local farming communities can be seen in the Egyptian property market back in the 1850s. Describing the courts that enforced the contracts of sale as 'a machine for transferring the land' from small farmers to the wealthy,[1] was a true description of the machinery required to police the system of new ownership of land as private property. Money suddenly became of vital importance. Farmers who had bought their ownership rights now had to mortgage their land in order to obtain loans to pay for costs arising out of their farming. All it took was one cattle epidemic and their creditors had the power to seize their land if they were unable to pay the debt.

---

1   Mitchell, T., *Rule of Experts: Egypt, Technopolitics, Modernity*, Berkeley, (2000), p 73.

Existing claims to the land under Ottoman and local law were codified, new courts were built, property registers were created, enforcement measures and complex repayment structures were put in place and banking became the primary source of credit for those who wished to use their property as collateral.[2] Investment from Europe began to pour in. New irrigation schemes in the countryside became an investment opportunity, new housing and infrastructure in the cities attracted a new breed of entrepreneurs. By the turn of the twentieth century the Egyptian stock market, whose largest share holdings were in mortgage companies and property development, was one of the major markets in the world. Meanwhile, small farmers faced rapidly rising prices.

The new leverage was debt. Colonial occupation of land was rife. The private property system was further consolidated with a land survey more comprehensive than anything known at that time in Britain.[3] The outcome was devastating. When a global depression struck in 1874, the Ottoman viceroy in Cairo foreclosed his extensive cotton and sugar cane estates; British and French banking houses stepped in to establish a Debt Commission. They took control of Egypt, her banks and her courts. They used the new courts to take possession of the viceroy's estates: when he objected to the takeover, the British and French governments intervened and removed him.

The subsequent rise of a constitutionalist movement governed by the army and a few rogue players provoked a British invasion in 1882 that ensured Europe had control over both finances and mortgaged property, including the extensive viceregal estates. Attempts to reduce the rate at which people were losing their land and their homes to creditors proved futile. By the 1920s more than one third of the agricultural population in the Nile Delta had lost their land. For the local Egyptian, private property rights had altered existing power structures beyond recognition within just a couple of generations. The same was true when the enclosure of commons began in Europe's colonies and in Europe itself. And

2 Ibid., pp. 59–74.
3 Ibid., pp. 84–93.

it remains true today. A World Bank supported programme that issued 8.7 million land titles in Thailand beginning in 1984 paved the way for acquisitions of land by a new breed of speculators, creating a scheme which undermined the peoples' tenure of land and security which in turn caused widespread rural conflict.[4] In Thatcherite Britain, the privatisation of social housing produced the gentrification of working class housing estates which pushed out many and caused a division which became polarized by the increase in the homeless versus the home-owner.

Sweden has a policy of supporting financial incentives for small businesses that wish to invest in clean-energy and low-carbon projects. This is fostered by a belief that clean development is compatible with creating markets for their companies which will in turn benefit both their economy and their well-being. It is a country that looks to the longer outcome for their people and recognizes that investment in sound business practice early on will reap greater rewards later. Policy mechanisms include laws that close the door to unethical social and environmental practices, and promote sustainable development for future generations. The Swedish Environmental Code, 1998, Chapter 1, Objectives and area of application of the Environmental Code, sets out at Section 1:

> *The purpose of this Code is to promote sustainable development which will assure a healthy and sound environment for present and future generations. Such development will be based on recognition of the fact that nature is worthy of protection and that our right to modify and exploit nature carries with it a responsibility for wise management of natural resources.*

4   Leonard, R., and Narintarakul na Ayutthaya, K., *Taking Land from the Poor, Giving Land to the Rich*, Watershed, Bangkok, Nov. 2002 – Feb. 2003, pp. 14–25.

The Environmental Code shall be applied in such a way as to ensure that:

1. *human health and the environment are protected against damage and detriment, whether caused by pollutants or other impacts;*
2. *valuable natural and cultural environments are protected and preserved;*
3. *biological diversity is preserved;*
4. *the use of land, water and the physical environment in general is such as to secure a long term good management in ecological, social, cultural and economic terms; and*
5. *reuse and recycling, as well as other management of materials, raw materials and energy, are encouraged with a view to establishing and maintaining natural cycles.*[5]

Ethical considerations for animals have likewise now been included in recent Swedish law to prohibit ownership by a person who has been found by the county administrative board to have maltreated, neglected or been convicted of cruelty to animals.[6] Similar animal protection laws are being advanced in various EU countries, with criminal sanctions. What this demonstrates is an indirect expansion of animal rights with criminal sanctions attached, namely the animal right to a healthy and cruel-free life. Animal rights are an extension of our concern and moral obligations which begin with self and family and extend out in ever expanding awareness to other people and races, then to other species and ultimately the land itself. This is known as an ever expanding circle of concern. Today's society is rapidly evolving to accommodate the extension of this circle of concern to the whole of the earth. Law merely affirms what is already becoming widely accepted within societal norms.

As a business, which may have tangential interests that extend into engagement with animals/nature/the environment,

---

5   Swedish Environmental Code, 1998, Chapter 1, s.1 http://www.sweden.gov.se/sb/d/2023/a/22847
6   Animal Welfare Act, consolidated 2010, s.29 (3) http://www.sweden.gov.se/sb/d/10158/a/90310

development of law in this area will have a direct impact on your practices and will shape the sphere of activity which will no longer be acceptable in the near future. New laws, which address environmental concerns, will arrive sooner than you may have factored in to your pre-existing business plans. The debate around what is required is rapidly evolving and taking giant strides. This is an arena that will arrive so speedily that your business may have to play catch-up. This book is designed to assist you in developing your company policy to accommodate the changes which will be arriving within the next few years. We live in rapidly evolving times where recognition of the damage and destruction to the earth is now reaching a level where it cannot be sustained and existing laws are found to be insufficient to prevent the destruction of ever expanding tracts of land and water. New laws, such as, criminalizing ecocide are just the beginning. It is a moment in history where the narrative around civilization's ability to adapt or die has woken to the reality that some hard and fast rules have to be put in place. Nowhere on Earth can be safeguarded from the onslaught of environmental destruction unless international laws with criminal sanctions which prohibit the destruction of the planet are now put in place.

This may seem draconian, but the consequences of allowing business to continue as usual are too disastrous to countenance. Trucost, a British company tasked with number-crunching for the United Nations Economics of Ecosystems and Biodiversity report in 2010, placed a conservative estimate on the damage and destruction to ecosystems by the top 3,000 corporations at $2.2 trillion for the year 2008.[7] This figure is exceeded by the GDP of only seven nations – a sobering message. Indications suggest that this trajectory is not slowing down. On the contrary it is speeding up. Somewhere along the line we have gone too far in the wrong direction and we are now escalating to such giddy heights of damage and destruction that biodiversity is threatened with reaching a global tipping point. We are about to reach a threshold

---

7   Jowitt, J., '3,000 Companies Cause $2.2 Trillion in Environmental Damage - Every Year', *The Guardian*, 18.02.2010

beyond which biodiversity loss will become irreversible, and all indications point towards us crossing that threshold within the next ten years if we do not make proactive efforts for conserving biodiversity.[8] Waiting for the moment to arrive will be too late. Time to act is now. That means your business shouldering it's environmental responsibilities and changing any destructive practices preferably before they are outlawed. This may be a quantum leap for some companies, but nothing short of a rapid turnaround will prevent the global implosion that will follow if we fail to act. This is our wake-up call.

BOOM AND BUST

Privatisation of property redistributes assets which in turn favours the wealthy over the poor. Extending intellectual private property rights over biological assets or land to those whose 'political resources are not commensurate with their new-found economic resources'[9] can all too often result in damaging, not improving, the lives of those who live on the land. Sooner or later the social fabric will begin to crack. Soon after 1991 when the Mexican government had passed a reform law to privatise the *ejido* lands, 'divesting itself of its responsibilities to maintain the basis' for indigenous security, the Zapatista rebellion broke out. Likewise Argentinian privatisation resulted in a 'huge inflow of over-accumulated capital and a substantial boom in asset values, followed by a collapse into massive impoverishment'.[10]

Creating ownership over land for its carbon-cycling capacity likewise brings all the attendant problems we have seen happening time and again throughout the world. It is merely another form of property trading which will engender enormous social change, impose rights over the atmosphere and fetter the earth's ability to regulate its own climate. It has already transformed or reinforced

8   Global Biodiversity Outlook report (2010) http://gbo3.cbd.int/the-outlook/gbo3/biodiversity-in-2010.aspx

9   Dove, M., Centre, *Periphery and Biodiversity: A Paradox of Governance and a Developmental Challenge*, Brush and Stabinsky, *Valuing Local Knowledge: Indigenous People and Intellectual Property Rights*, Island Press, Washington, DC (1996) pp. 41–67

10 Harvey, D., *The New Imperialism*, Oxford University Press (2003) p. 158

a wide range of power relations – by, for example, creating new institutions to quantify, handle, regulate, distribute and police the new assets that are being traded or given away. Turning the earth's carbon-cycling capacity into a tradable asset is already proving to be a false solution.

BUSINESS RESCUE PACKAGE

Most countries have business rescue legislation. Clearly, the ability to rescue economically viable companies experiencing financial difficulties is in the best interests of shareholders, creditors, employees and other stakeholders, if in the interests of the country as a whole. Take this a step further; to rescue economically viable companies makes sense where their business is good for the earth. A business that places the earth at risk, however, also places humans lives at risk. It may seem inconsequential to have one business destroy a small patch of land, but multiply the practice and increase the number of companies doing the same throughout the world and the impact on the earth is considerable. Companies that are functioning from a position of limiting their impact are ensuring that human well-being is prioritised over profit. Companies that are destroying the earth do not have a well-being provision. So there is a question of which businesses can remain functioning for the greater success of the global ecology, not just the global economy. Shore up a company that is plundering the earth and we are left with a world where damage and destruction has become so normalised that it becomes impossible to break the hermetically sealed spiral of escalating ecocide.

All nations have the power to pass emergency laws overnight: that's often how emergency business rescue packages are put in place. Wall Street was saved by last-minute intervention; terrorist legislation was passed within days in the UK and many other countries use similar provisions when in times of *extremis*. Banks and businesses have been subject to last minute bail-outs from governments. The earth, unlike a bank, has no price tag. However we know that this is one entity that is in great need of an emergency rescue package. Laws can be passed overnight to prevent further collapse of its ecosystems, and toxic assets can be remedied. This is one bail-out that would change the world and all businesses that are dependent on it functioning. Set in place the right emergency rescue package, and the world will benefit.

DANGEROUS INDUSTRIAL ACTIVITY

When crude oil and gas is extracted from the ground, it is always mixed with an amount of water.

Produced water from offshore drilling operations is highly dangerous; as wells are drilled deeper and deeper every year the water becomes increasingly hazardous, toxic, even radioactive. This issue has been hidden from sight by all oil companies throughout the entire world, with the exception of Norway. Most highly toxic produced water is simply separated out from the crude oil or natural gas right on the offshore platforms and immediately dumped into the sea, usually through caisson pipes that discharge about fifty meters below the sea's surface. Properly detoxifying produced water is a complicated and expensive business, so the answer is to simply and quietly dump it offshore. No-one queries it and it goes under the radar. Each time produced water is dumped, the seas suffer and the discharge is left to disperse.

THE RIGHT TO POLLUTE

More commonly known property rights include freeholds, leaseholds, licenses, patents, easements, quotas, copyrights, concessions, and usufructs. Written and verbal rights, also known as formal rights and informal rights, are sometimes established over time out of ancient common rights, of which there are hundreds of kinds. They overlap, influence and evolve as we interact with our land. Different groups can all at the same time view a plot of land as private, public and common property. People have invented property rights of many different kinds. In order to use, defend, steal or appropriate the things they want and need, a whole body of law has come into being to make this happen. Property laws are no longer restricted to land, but also now extend to property rights governing everything from land and water to birds' nests, the radio spectrum, ideas and DNA. There are rights to exclude, to use, to benefit from, to inherit, to manage, to transfer. There are rights that are held by communities, rights that are held by individuals, and rights that are held by the state. Private property is guaranteed by, but subject to, the authority of the state and the public: individual user rights of commoners tend to be granted at the will of the community.

The British jurist Sir Henry Maine, a century and a half ago, compared different kinds of property systems to different 'bundles of sticks.' With these bundles, one could include the right to pass them on to your heirs – or not, depending on your whim. Some bundles include the right to buy and sell, some do not. Other rights could be added as one so wished: the right to use, to have access to, to manage, to exclude etc. As political scientist Elinor Ostrom notes: 'None of these rights is strictly necessary... Even if one or more sticks are missing, someone may still be said to 'own' property... one must... specify just what rights and corresponding duties [a] regime would entail.'[11]

Tradable pollution allowances and credits are at the end of the day property and thereby create rights of ownership for those who buy, use and trade them. They are another way of supporting 'enforceable claims to use something': to pour carbon dioxide into the oceans, soil and vegetation. To make money through trading allowances, or to gain a competitive advantage by procuring carbon permits that others have to pay for, guarantees for the user/owner the 'enforceable rights to benefit from something'.[12] Pollution allowances and credits also contain benefits that include tradability, 'excludability' – for example, Scottish Power cannot use Ineos Fluor's allowances or credits[13] and credibility. To say that emissions allowances are not property rights – meaning merely

11 Elinor Ostrom, *Governing the Commons: The Evolution of Institutions for Collective Action*, Cambridge University Press, Cambridge 1990.
12 Eds: Lohman et all, 'Carbon Trading, a Critical Conversation on Climate Change, Privatisation and Power', *What Next*, issue No. 48 (2006) p.76
13 Ibid.

that they are not permanent – is, without doubt, a deception of enormous magnitude.

NATURE HAS NO PRICE

There is a growing momentum for Earth rights to be endorsed by the United Nations. Countries best known for their indigenous wisdom are beginning to speak out. Resolutions such as the 'Harmony with Mother Earth' (A/ C.2/64/L.24) was co-sponsored by Algeria, Benin, Belarus, Bosnia and Herzegovina, Cape Verde, Cuba, Dominica, Ecuador, Eritrea, Georgia, Guatemala, Honduras, Mauritius, Nepal, Nicaragua, Paraguay, Saint Vincent and the Grenadines, Saint Lucia, Seychelles and Venezuela.

Speaking to the packed hall, Mr Solon stated to the UN: 'We acknowledge and share the progress of the environmental agenda of the United Nations at the level of the biodiversity, the ozone layer, desertification, climate change and other sectors, but we are convinced that this needs to be supplemented with a more holistic approach given the serious global impacts we are witnessing.'

Ministers from Cuba, Ecuador, Nicaragua and Venzuela, Authorities of the Ministerial Committee for the Defence of Nature of the Plurinational State of Bolivia and members of the Bolivarian Alliance for the Americas gathered in La Paz, Bolivia from 3–5 November 2010. The Bolivian Minister of the Environment and Water, Maria Esther Udaeta and environment ministers of Ecuador, Cuba, Nicaragua, and Venezuela all participated in drafting the constitution of the new Ministerial Committee on the Defense of Nature.

Ecuador is the most advanced of all nations in recognizing nature's rights. As a signatory of the Declaration of the Ministerial Committee for the Defense of Nature of ALBA-TCP, they endorsed on 5 November 2010 the proclamation that 'Nature has no price'.

Along with the Bolivian Minister of the Environment and Water, Maria Esther Udaeta, and fellow environment ministers of Cuba, Nicaragua and Venezuela, they are now drafting the constitution for the new Ministerial Committee on the Defense of Nature. Ecuador is the first country to implement laws to protect nature's rights, and the first country to accommodate the filing of a suit against BP and its crimes against nature in their Constitutional Court from an international coalition of defenders of nature's rights.

. . . .

DECLARATION OF THE MINISTERIAL COMMITTEE FOR THE DEFENSE OF NATURE OF ALBA-TCP

We declare:

1. That nature is our home and is the system of which we form a part, and that therefore it has infinite value, but does not have a price and is not for sale.

2. Our commitment to preventing capitalism from continuing to expand in the spheres that are essential to life and nature, being that this is one of the greatest challenges confronting humanity.

3. Our absolute rejection of the privatization, monetization and mercantilization of nature, for it leads to a greater imbalance in the environment and goes against our ethical principles.

4. Our condemnation of unsustainable models of economic growth that are created at the expense of our resources and the sovereignty of our peoples.

5. Only a humanity that is conscious of its present and future responsibilities, and states with the political will to carry out their role, can change the course of history and restore equilibrium in nature and life as a whole.

6. That instead of promoting the privatization of goods and services that come from nature, it is essential to recognize that these have a collective character, and, as such, should be conserved as public goods, respecting the sovereignty of states.

7. It is not the invisible hand of the market that will allow for the recuperation of equilibrium on Mother Earth. Only with the conscious intervention of state and society through policies, public regulations, and the strengthening of public services can the equilibrium of nature be restored.

8. Cancun cannot be another Copenhagen; we hope that accords will be reached in which developed countries truly act according to the principle of common but differentiated responsibilities, and effectively assume their obligation to reduce greenhouse gas emissions, without making climate change into a business through the promotion and creation of carbon market mechanisms.

9. That, committed to life, the countries present at this meeting agree to include in our permanent agenda, among other actions, the realization of a referendum on climate change and the promotion of the participation of the peoples of the world.

10. That it is urgent to adopt at the United Nations a Universal Declaration of the Rights of Mother Earth.

THE LAW OF ECOCIDE AS A TOOL TO RAISE OUR COLLECTIVE CONSCIOUSNESS

Shifts in consciousness are rarely gradual. Evidence demonstrates that when pressure mounts to a certain point, human evolution experiences a jump in depth of understanding. At that juncture a shift takes place, a shift predicated on the recognition of an intrinsic value. That shift is then manifested in words and action. Examples of such shifts in the history of civilization include the point of recognition that the world was not flat, the abolition of slavery, the collapse of the Berlin Wall, the end of apartheid. We are at a pivotal point in the whole of the history of civilization. We are waking up to the realisation that we are about to make a similar shift. We know now that we can end our damage and destruction of our Earth. All we need do is change the laws to shift our current premise to one where caring for the earth comes first.

Part 2

SHIFTING THE PARADIGM

Your life is what your thoughts make it

Chapter 4

BUILDING A NEW BUSINESS MODEL

E COCIDE straddles both types of nuisance – public and private. The wording of the definition of ecocide is deliberate. It is designed to incorporate both civil and criminal legislation that will grow out of the definition. Prosecution for the crime of ecocide brings with it civil litigation as well. Restoration of damage, destruction or loss of ecosystems will be a key aspect to be determined in the light of convictions being made. National legislation premised on the ability of a person or group of persons to bring a claim in negligence, against a corporate entity whose directors have been found guilty of ecocide can be determined speedily and at limited cost. In such circumstances instead of the claimant having to prove his case, the conviction of the CEO and/ or his directors will take the case straight to the issue of damages and compensation to be payable by the company. Thus, a CEO can be imprisoned for ecocide activities (and restoration of damage ordered as part of the sentencing requirements), plus the inhabitants of the territory could take individual or class action against the company for compensation.

BOLIVIA'S NEW EARTH LAW

A new law, *La Ley de Derechos de la Madre Tierra*, or 'The Law of Mother Earth' has just received approval by the people of the Plurinational State of Bolivia. It is an historic moment: in granting rights to the earth the traditional but silent corporate rights – the right to extract, the right to pollute and the right to destroy – will no longer be the primary drivers of business decisions. How this plays out in their political and corporate arena remains to be seen; Bolivia earns $500m (£305m) a year from mining companies which provides nearly one third of the country's foreign currency. Earth Rights will prevent mining and heavy extractive businesses from continuing in the fashion they have been used to. Instead they would be subject to restoration, practices which ensure the land is not destroyed – or where it has been, it will be remedied with the best methods for the health and well-being of the earth. Where it is deemed excessive to mine or extract from the land, without restoring the balance and harmony of the ecosystems within a short timescale, the extraction will be deemed illegal.

Future obligations on both the government and industry to halt destructive and damaging activities will now come under scrutiny as the world looks on. What is clear is that Bolivia will have to balance these environmental imperatives against industries – like mining – that contribute to the country's GDP.

LAW OF MOTHER EARTH

In December 2010 the Bolivian Government implemented a draft Law of Mother Earth, to establish new rights for nature. They include the right to not have cellular structure modified or genetically altered; the right to continue vital cycles and processes free from human alteration; the right to pure water; the right to clean air; the right to be free of toxic and radioactive pollution; the right not to be affected by mega-infrastructure and development projects that affect the balance of ecosystems and the local inhabitant communities.

The draft of the new law states: 'She is sacred, fertile and the source of life that feeds and cares for all living beings in her womb. She is in permanent balance, harmony and communication with the cosmos. She is comprised of all ecosystems and living beings, and their self-organisation.' Indigenous belief is that 'Pachamama' is a living being, not a mere thing.

Decrees: Act of the Rights of Mother Earth
Chapter I - Object and principles
Article 1. (Scope).

This Act is intended to recognize the rights of Mother Earth, and the obligations and duties of the Plurinational State and society to ensure respect for these rights.

Article 2. (Principles).

The binding principles that govern this law are:

1. **Harmony.** Human activities, within the framework of plurality and diversity, should achieve a dynamic balance with the cycles and processes inherent in Mother Earth.

2. **Collective good.** The interests of society, within the framework of the rights of Mother Earth, prevail in all human activities and any acquired right.

3. **Guarantee of the regeneration of Mother Earth.** The state, at its various levels, and society, in harmony with the common interest, must ensure the necessary conditions in order that the diverse living systems of Mother Earth may absorb damage, adapt to shocks, and regenerate without significantly altering their structural and functional characteristics, recognizing that living systems are limited in their ability to regenerate, and that humans are limited in their ability to undo their actions.

4. **Respect and defend the rights of Mother Earth.** The State and any individual or collective person must respect, protect and guarantee the rights of Mother Earth for the well-being of current and future generations.

5. **No commercialism.** Neither living systems nor processes that sustain them may be commercialized, nor serve as anyone's private property.

6. **Multiculturalism.** The exercise of the rights of Mother Earth requires the recognition, recovery, respect, protection, and dialogue of the diversity of feelings, values, knowledge, skills, practices, skills, transcendence, transformation, science, technology and standards, of all the cultures of the world who seek to live in harmony with nature.

Chapter II - Mother Earth, definition and character
Article 3. (Mother Earth).

Mother Earth is a dynamic living system comprising an indivisible community of all living systems and living organisms, interrelated, interdependent and complementary, which share a common destiny.

Mother Earth is considered sacred, from the worldviews of nations and peasant indigenous peoples.

Article 4. (Living Systems).

Living systems are complex and dynamic communities of plants, animals, microorganisms and other beings and their environment, where human communities and the rest of nature interact as a functional unit under the influence of climatic, physiographic, and geological factors, as well as production practices, Bolivian cultural diversity, and the worldviews of nations, original indigenous peoples, and intercultural and Afro-Bolivian communities.

Article 5. (Legal status of Mother Earth).

For the purpose of protecting and enforcing its rights, Mother Earth takes on the character of collective public interest. Mother Earth and all its components, including human communities, are entitled to all the inherent rights recognized in this Law. The exercise of the rights of Mother Earth will take into account the specificities and particularities of its various components. The rights under this Act shall not limit the existence of other rights of Mother Earth.

Article 6. (Exercise of the Rights of Mother Earth).

All Bolivians, to join the community of beings comprising Mother Earth, exercise rights under this Act, in a way that is consistent with their individual and collective rights.

The exercise of individual rights is limited by the exercise of collective rights in the living systems of Mother Earth. Any conflict of rights must be resolved in ways that do not irreversibly affect the functionality of living systems.

Chapter III - Rights of Mother Earth

Article 7. (Rights of Mother Earth)

I. Mother Earth has the following rights:

1. **To life:** The right to maintain the integrity of living systems and natural processes that sustain them, and capacities and conditions for regeneration.

2. **To the diversity of life:** It is the right to preservation of differentiation and variety of beings that make up Mother Earth, without being genetically altered or structurally modified in an artificial way, so that their existence, functioning or future potential would be threatened.

3. **To water:** The right to preserve the functionality of the water cycle, its existence in the quantity and quality needed to sustain living systems, and its protection from pollution for the reproduction of the life of Mother Earth and all its components.

4. **To clean air:** The right to preserve the quality and composition of air for sustaining living systems and its protection from pollution, for the reproduction of the life of Mother Earth and all its components.

5. **To equilibrium:** The right to maintenance or restoration of the interrelationship, interdependence, complementarity and functionality of the components of Mother Earth in a balanced way for the continuation of their cycles and reproduction of their vital processes.

6. **To restoration:** The right to timely and effective restoration of living systems affected by human activities directly or indirectly.

7. **To pollution-free living:** The right to the preservation of any of Mother Earth's components from contamination, as well as toxic and radioactive waste generated by human activities.

Chapter IV – State obligations and societal duties

Article 8. (Obligations of the Plurinational State).

The Plurinational State, at all levels and geographical areas and across all authorities and institutions, has the following duties:

1. Develop public policies and systematic actions of prevention, early warning, protection, and precaution in

order to prevent human activities causing the extinction of living populations, the alteration of the cycles and processes that ensure life, or the destruction of livelihoods, including cultural systems that are part of Mother Earth.

2. Develop balanced forms of production and patterns of consumption to satisfy the needs of the Bolivian people to live well, while safeguarding the regenerative capacity and integrity of the cycles, processes and vital balance of Mother Earth.

3. Develop policies to protect Mother Earth from the multinational and international scope of the exploitation of its components, from the commodification of living systems or the processes that support them, and from the structural causes and effects of global climate change.

4. Develop policies to ensure long-term energy sovereignty, increased efficiency and the gradual incorporation of clean and renewable alternative sources into the energy matrix.

5. Demand international recognition of environmental debt through the financing and transfer of clean technologies that are effective and compatible with the rights of Mother Earth, among other mechanisms.

6. Promote peace and the elimination of all nuclear, chemical, and biological arms and weapons of mass destruction.

7. Promote the growth and recognition of rights of Mother Earth in multilateral, regional and bilateral international relations.

Article 9. (Duties of the People)

The duties of natural persons and public or private legal entities:

1. Uphold and respect the rights of Mother Earth.

2. Promote harmony with Mother Earth in all areas of its relationship with other human communities and the rest of nature in living systems.

3. Participate actively, individually or collectively, in generating proposals designed to respect and defend the rights of Mother Earth.

4.   Assume production practices and consumer behaviour in harmony with the rights of Mother Earth.
5.   Ensure the sustainable use of Mother Earth's components.
6.   Report any act that violates the rights of Mother Earth, living systems, and/or their components.
7.   Attend the convention of competent authorities or organized civil society to implement measures aimed at preserving and/or protecting Mother Earth.

Article 10. (Defense of Mother Earth).
Establishing the Office of Mother Earth, whose mission is to ensure the validity, promotion, distribution and compliance of the rights of Mother Earth established in this Act. A special law will establish its structure, function, and attributes.

....

Permaculture has a fundamental principle which applies to all of life: *the problem is the solution.* The problem for corporations is a) the law that governs what is prioritised and b) the prevailing corporate structure that places profit before all other considerations. Both can change – and can change with ease. All it takes is change to the laws and change to the prevailing practices. One makes it easy for the other to change; without change in the law, 'business as usual' is a far easier option to retain. It takes determined decision-making by corporations to change their culture without the assistance of law. This is one of the reasons why we have so much talk without action. It actually proves counter-productive to encourage voluntary action, as the majority of businesses can never move as fast as they would wish to evolve into the companies they espouse to be. Law can shift corporate activity on a pin-head. However, where the corporate thinking remains stuck in a pre-set mould, the friction can prove to be counter-productive. Both are required to change at the same time if success is to be the likely outcome.

Making change can be incredibly rewarding and challenging. It can also be frightening to some and that can prevent productive and pre-emptive action from being taken. Quite literally a corporate culture can become caught like rabbits in the headlights whilst the crash looms like an enormous shadow above. Too often in these scenarios the first the public hear of it is when it hits the media that enormous losses have been incurred. Simple steps taken at earlier junctures could lead to very different outcomes. A decision to embed new values that are driven by environmental and ethical concerns will, for instance, prevent investment into unsustainable markets which will in time lead to collapse. Thinking long-term is rarely undertaken by corporations whose primary concern is quarterly growth margins. Looking forward at the impact of a given outcome to the next generation is rarely considered and there are no laws in place to govern business in this way. Yet, almost all business impacts on future generations in a myriad of ways, from waste legacy to unresolved landfill issues. Health and well-being of both humans and the environment are the primary legacy we leave behind. At the moment the norm is an adverse impact – not a positive one – that is being carried forward.

In War and Peace

Crimes Against Peace, as the four crimes are known – crimes against humanity, war crimes, crimes of aggression and genocide – are recognised as international crimes that lead to conflict. They are adopted under all War Codes and are placed at the top of all nations 'to avoid' lists. However, what is left out of the list is ecocide. Where there are limited resources that are at risk of being lost, conflict can easily escalate and where conflict arises resource wars can follow. It becomes a hermetically sealed cycle of damage and destruction leading to resource depletion leading to conflict and sometimes war. Never before have we faced such an onslaught on nature's resources as we do now.

Sudan's peace accord, signed in 2005 after a toll of over two million people, was built around agreements governing its natural resources. It was understood that without determining resource distribution, peace would never prevail. Thus, one of the core agreements was to divide oil revenues equally between the government and the rebels. More often than not peace negotiations ignore the issue altogether. A recent UN report that reviews data on conflict, observed that over the past sixty years 'fewer than a quarter of peace negotiations aiming to resolve conflicts linked to natural resources have addressed resource management mechanisms'. In economic terms the fight over natural resources, in particular land and raw materials, is both a cause of war and a source for financing conflict. Natural resources are inevitably an important source of power for state and non-state armed groups as recent investigations by both the UN Panel of Experts and Global Witness suggest. Peace processes currently have no means of addressing this issue and rebel commanders are using their control over cocoa, diamonds and gold to finance renewed weapons purchases. Strategies for shifting an armed conflict towards peace remains focused on the warring parties' political and military capabilities. However, it fails to take account of the control of natural resources which inevitably raises the risk of a return to armed violence.

Where new-found stability is agreed, lack of proper determination of natural resources can be a recipe for future conflict. Post-peace agreement, as Liberia's transitional administration has experienced in the past five years, the legacy of a corrupt political and economic system, remain to be unravelled by any democratic government that is subsequently elected. As reported by the international NGO Global Witness, a number of Western diplomats admitted that they and others have been reluctant to discuss the issue of natural resources with the governments of the Democratic Republic of Congo, Rwanda and other neighbouring countries, because they consider it too sensitive. One UN official stated that 'natural resources are not on the table of topics in peace talks. Almost every other issue is. Yet it's one of the keys to resolution of the conflict. The peace talks discussed the framework for the army, brassage (intermingling, part of the integration process), demobilisation, etc but not natural resources. Yet the armed groups are not prepared to leave the resources behind.'[1]

---

1   Global Witness interview with MONUC official, Goma, July 2008, *Lessons Unlearned: How the UN and Member States must do more to end natural resource-fuelled conflicts*, Global Witness (2010) p.18

GLOBAL WITNESS RECOMMENDATIONS

Global Witness recommends that peacemakers should mandate a team to report on the implementation of due diligence measures by companies active in, or sourcing materials from, conflict zones. In their report, *Lessons Unlearned*, various recommendations pertaining to natural resources are set out. They include the following:

Peacemakers should approach the natural resource dimensions of conflict resolution with a view to setting 'the rules of the game' governing the parties' economic activities during the peace process. This is an important foundation for both peacekeeping and peace-building efforts.

- Address the economic interests of warring parties as a central part of the overall approach to conflict resolution.
- Seek to demilitarise control of natural resources.
- Establish the 'rules of the game' for the transition of the war economy to a peace-building economy.
- Avoid deals which 'lock in' poor governance of natural resources. Limits should be placed on the ability of unelected transitional governments to allocate natural resource concession contracts.
- Build an independent monitoring mechanism into any natural resource wealth-sharing provisions of a peace agreement.
- Incorporate a dispute resolution mechanism into any natural resource wealth-sharing provisions of a peace agreement. This might consist of an agreement to refer disputes to an arbitration tribunal.
- Require international guarantors of a peace agreement to play a role in enforcing any provisions concerning natural resource management. The cost to the parties of a failure to adhere to these provisions should be clear, significant and enforceable by law.
- Ensure that any attempt to bring informal and illegal activities relating to natural resources into the formal economy should be based on clear and verifiable standards of behaviour backed by regulation.

- Draw on natural resource assessments by experts, such as those on the UN's Mediation Support Unit roster, that provide a description of what resources are at issue, their potential values and their relevance to the negotiations.

Global Witness believes that peacekeepers intervening in conflicts that have a natural resource dimension must be mandated to deal with it directly. Peacekeepers should be authorised to work with local, national and regional customs and law enforcement officials, and international monitors or panels of experts, to investigate, to monitor trade routes and border crossings, and to assist in inspections by customs and other government officials. Peacekeeping operations should also be authorised to intercept illicit trade that supports armed groups, while recognising that doing so will depend on the capacities of the mission and the tactical situation on the ground.

Global Witness is recommending that the UN Security Council should:

- Mandate peacekeeping operations to respond to the natural resource dimensions of conflicts.
- Request that the Department of Peacekeeping Operations establish operational guidelines for peacekeepers on how to respond to the problem of illicit natural resource exploitation and trade in the theatre of operations.
- Authorise peacekeeping missions to enforce sanctions and laws governing the exploitation and trading of natural resources, including the interdiction and confiscation of shipments, where operational considerations allow.
- Authorise peacekeepers to deploy to sites of natural resource production, where operational considerations allow, to protect these sites from exploitation by abusive state or non-state armed groups, and to protect people living and working at these sites.
- Require cooperation between joint mission analysis cells, expert panels, and local and regional customs and law enforcement agencies to track and intercept shipments of conflict resources.

- Authorise peacekeepers to deploy to protect those international and local officials seeking to police the exploitation and trading of natural resources.[*]

---

[*] *Lessons Unlearned: How the UN and Member States must do more to end natural resource-fuelled conflicts*, Global Witness (2010), pp.43 - 47

Chapter 5

THE FLOW OF MONEY

IT's all about the flow of money. Money flows into damage and destruction every day. We have created a system that gives legal credence to making money out of businesses that create ecocide. Ownership gives the right to do as we wish. Once the contract is signed and the ink is dry, money is to be made by destroying the land beneath us – or if not directly beneath us then beneath others' whose rights have been over-ridden. Markets are created to capture the profit and banks provide the capital outlay to make it all happen.

Profit per se is not a problem; it becomes a problem when it is generated from damaging and destructive activity. Where ecocide is occurring on a daily basis, the norm has been set to a level that has become untenable. Making money out of businesses that pay little regard for the consequences of their decision-making is liable to result in an escalation of current practices. Without laws in place to halt the banks and financial institutions from financing ecocide, money flows to the biggest ecocides of all.

Where money flows, businesses are built. Many innovative solutions fail without proper investment: deemed too high-risk, they are shelved until new laws create a fresh impetus to financing innovation to replace the old. Law acts as the lever to change the direction of the train: as it reaches a junction, the lever shifts the points at the very last minute – so too with laws that are brought in when we reach crisis point.

Money can flow into solutions that help ease the pressure on the world to prevent carbon displacement. Forests are an obvious target as they provide invaluable carbon sinks. However, creating ownership of the forests to traders has proven to be a flawed solution. Indigenous rights are marginalised by the silent corporate right to own and trade. Reducing Emissions from Deforestation and Forest Degradation (REDD) is an effort to create a financial value for the carbon stored in forests by offering incentives for developing countries to reduce emissions from forested lands and invest in low-carbon paths to sustainable development. 'REDD+' is the second attempt to extend the remit by extending trading rights over land management. Where colonisation is defined as plundering of people and planet, REDD+ facilitates the flourishing of eco-colonisation of the forests. Compliance and non-compliance of engagement with local forest communities is not addressed: it is relegated to a 'safeguard' for use if so inclined, rather than being obligatory. Whilst the annex notes that the UN General Assembly has adopted the Universal Declaration on the Rights of Indigenous Peoples, the REDD+ text simply refers to indigenous peoples' rights. Somehow, the REDD+ document has managed to relegate Indigenous rights to nothing less than a passing whim. Crucially there is no specific leverage to protect their rights. REDD+ is a document without teeth, a limp pronouncement of the legally established rights of those who live upon the land that is now termed 'carbon stock.' In one fell swoop their rights have been consigned to the margins. The silent right to trade has trumped indigenous rights in the fight to save the planet. The interests that are being protected are those of the businesses that stand to profit by the deal. Neither the land nor the people, whose territory is being traded in a market of potentially massive proportions, are being protected by REDD+.

IT'S NOT WORKING

'Despite the increased financial commitment to renewable energy sources and successful management and utilisation of carbon finance, there are serious shortfalls in the World Bank's investment framework.' So concluded the Swedish Society for Nature Conservation who published a comprehensive report in response to the World Bank's Clean Energy for Development Investment Framework. They noted that the World Bank is at substantial risk of missing the window of opportunity to pave a genuine pathway to low-carbon and climate resilient development. Firstly, their investment portfolio still favours large-scale hydropower projects and/or uncertain and expensive technological options. Considerable resources are still allocated to fossil-fuel based installations and projects. Secondly, the dominance of large-scale projects and large project developers is still pervasive in the Clean Development Mechanism (CDM). The CDM does not necessarily guarantee the accessibility of the carbon market to Small and Medium-sized Enterprises (SMEs) who are more active and innovative in renewable technology markets. Thirdly, while the role of sectoral reforms and private participation are important for mobilising private investment, 'private cash-flow-driven strategy' does not work well in low-income and basic-need dominated markets. Energy policy reforms supported by the World Bank should integrate with market liberalisation and sustainable initiatives in Africa and link the power access targets with potential targets for both renewable energy and sustainable development.*

---

* Nannan Lundin, N., and Hagberg, L., *Report: An Assessment of the World Bank's Clean Energy for Development Investment Framework*, Swedish Society for Nature Conservation

....

World Bank Assessment Rules
Put in place a law that prohibits an act and all other laws must be brought into line. When this happens, the trickle-down effect is immediate: banks come first. Where money is directed determines whether finance is available for an activity or project. Close the door, change the rules and the money shifts elsewhere. Close the door to ecocide, make it a crime and the rules for banks change. In Appendice III *World Bank Assessment Rules (as amended)* are included as best practise for all investment purposes. The World Bank Group report, *Development and*

Climate Change, A Strategic Framework for the World Bank Group (2008), states they will make a

> 'dedicated effort to grow resources and capacity for supporting resilience to climate risks and adaptation efforts, and work towards closing the gap in the vastly different levels of knowledge, experience and financial resources currently available for adaptation and mitigation.'

Its first stated objective is to enable the WBG to support sustainable development and poverty reduction effectively at the national, regional, and local levels, as additional climate risks and climate-related economic opportunities arise. Climate Investment Funds (CIF) have been set up to do this. Its second objective is to use the WBG's potential to facilitate global action and interactions by all countries to mitigate against climate change.[1] The only way to achieve these two objectives is to close the door to financing damaging and destructive projects.

The World Bank brought out their *Environmental Assessment Rules* in 1999 for assessing projects that have environmental impact. Their rules did not change despite the strong, stated objectives in their 2008 report. It is one thing to state their intent, it is another to follow it through. To do that, the rules have to change.

The first rule to be changed is to prohibit financing of projects that run counter to climate mitigation. The *World Bank Assessment Rules* have been updated here to prevent financing of Category A projects, projects that cause environmental impact on a scale that is at risk of causing ecocide. Each time a project is submitted, a decision is made as to whether it is one of three categories; A, B or C. Under the Amended Rules A is no longer legal, and any project that comes under it's ambit will be referred for consideration of prosecution. Category B will now be subject to more scrutiny and will be opened up for public consultation, and NGO and expert input, before a final decision is made.

---

1 *Development and Climate Change, A Strategic Framework for the World Bank Group*, 2008, see also http://climatechange.worldbank.org/climatechange for interim reports.

These are rules which can be applied across the world; they needn't be retained solely for the purpose of World Bank projects and they are designed to be best practise for all banks, investment funds and other related bodies. The rules are designed for all projects, large and small, and the rules can be used at a local and regional level as well. By making the rules for the world's largest and most powerful bank the benchmark for all others, banks throughout the world have a level playing field within which to play. Financing of ecocide projects will become the exception, not the norm and finance will flow readily into environmentally benign projects instead. By altering the flow of money the new green economy gets the green light, policymakers can begin to pave the way to a clean energy future and jobs are created.

By making banks and other financial institutions refuse finance to the major polluters, investment signals are dramatically altered: financiers will have surety of investment and long-term returns. Companies that were previously hindered from changing direction by lack of investor support will find more willingness in the market to invest in new and innovative projects that are Category B and C. Policy makers will receive signals from the market that they need to push further with plans to advance new technological solutions and take them to mainstream level with the help of new laws and subsidies.

In creating the fit-for-purpose framework for banks, the market-place changes dramatically. Rather like inserting new software into a machine, the cogs whirr and speed up; new banking rules do just that. Identifying what can and cannot be invested in sends out a very stable and clear message to the world. Companies work well with new legislation: it gives them the boundaries within which to operate. Banks with the same rules will find that they will attract the businesses that they want to invest in and will want to support businesses that are facing up to the challenge.

Businesses will be evaluated on their overall impact on the environment, not just on carbon emissions. There will be an implicit recognition that carbon markets are not the answer. Environmentally benign business brings with it a far smaller carbon footprint in any event. Prohibition is far easier to establish than any monitoring system which can always be subject to evasion or corruption. By closing the door to a level of project that is clearly defined in law as well as in best practice, creates a space for responsibility to grow. Banks will no longer want to finance damaging and destructive projects: they will have the easy response that it is a crime to do so. Investment of a Category A project will put all senior managers and directors at risk of prosecution, which acts as the most powerful disincentive of all. Very quickly it will become the norm to disassociate from and discourage any projects that could give rise to potential litigation for the bank. Where the rules apply worldwide, business will change overnight to reinventing their wheels as best they can. Governments will be required to help those who need assistance and if necessary provide emergency rescue packages for those who need the most assistance in changing their workforce skill-base. Subsidies for Category A projects will no longer exist, freeing up finance to help those who will be most needed to re-skill and re-tool. It is the route to making sure that the green economy flourishes, and governments will respond accordingly by helping those companies who take up the challenge.

IF ALLOWANCES ARE TEMPORARY, SURELY THEY HAVE NO PROPERTY RIGHTS?

It's not so simple. Just because something is temporary, does not mean it is not a property right.

Property rights come in many shapes and sizes. A lot of property rights are temporary. Think of monthly or yearly leases. Think of mining, logging or grazing concessions that governments give out to corporations for thirty years or seventy-five years. Think of copyrights, trademarks, and licenses. Think of fishing quotas or seed, gene or drug patents, all of which expire after a certain length of time.

All of these temporary property rights have been used to privatise or enclose various goods. All have been used to make billions for private companies. And all have been used to transfer wealth and power to the rich, sometimes igniting bitter conflict over democracy and how human beings' environments are to be treated.

Emissions allowances are no different. Industry, economists, governments and legal scholars all agree that, in giving away these allowances, emissions trading schemes do give away something quite substantial.

As the International Accounting Standards Board notes with regard to the EU's Emissions Trading Scheme, allowances are 'assets... owned by the company concerned... and as such represent a significant and immediate creation of value to companies'. They should be seen as a 'government grant, and accounted for as such, i.e. treated as deferred income in the balance sheet and recognised as income on a systematic basis'. Temporary or not, emissions permits constitute a 'major input factor to production.'

Allowances aren't valuable just because they enable polluters to avoid having to spend money on pollution control. They also enable corporations to borrow money more easily and give them a better share price. And they set a precedent for granting them further entitlements. They can also be bought and sold for clear profit. They have market value. It matters who they are given to.

CREATING PROPERTY OUT OF HOT AIR

Now we have created markets literally out of the air. Carbon markets are a fictional market based on the trading of the carbon that can be stored in the ground and in forests. It's akin to trading an idea on the basis that it has yet to become a reality; a bit like going to the bookies and putting money on a fictional character winning the race, only they have yet to be included in the programme. Before they are, the fictional character can be traded from person to person before being placed in the race. The thing is, they may never get to the race.

In many countries, governments have granted open-ended 'temporary' property rights that have become permanent in all but name. In many States 'temporary' commercial mining and logging concessions, leases and licenses which are valid on a twenty plus year contract end up with public or community land being handed over to the private sector for good. A typical result is where corporations have held onto their vast timber leaseholds by converting them to plantation crops or exploiting their minerals, often using old logging roads and dispossessing hundreds of thousands of local residents who have little access to the judicial system. This happens time and again in Indonesia.

HOW CAN YOU HAVE RIGHTS OVER SOMETHING AS INTANGIBLE AS THE EARTH'S CARBON-CYCLING CAPACITY?

Companies have legal rights over all sorts of intangible things. Drug companies own genes. The Disney Company owns the Winnie-the-Pooh story. General Electric and Rupert Murdoch hold temporary rights over parts of the broadcast spectrum – rights that they are now trying to make permanent. Other companies own new ideas for their production lines.*

---

*   Eds: Lohman et al, 'Carbon Trading, a Critical Conversation on Climate Change, Privatisation and Power', *What Next*, issue No. 48, 2006, p. 77

US grazing permits illustrate how distribution of permits that the government nominally retains 'control' over can in effect privatize a resource. Just as today's Kyoto Protocol and the EU's Emissions Trading Scheme (ETS) allowances are given to those who are the biggest polluters of the atmosphere, the Taylor Grazing Act of 1934 gave grazing permits to those who were the biggest land owners. The grazing permits could be both limited and revoked by the government who explicitly claimed they did not amount to 'rights, title, interest or estate in or to lands'. The number of permits were to be reviewed and periodically adjusted by the Department of the Interior who were tasked to regulate the rangelands. Congress viewed the permits as mere privileges and wanted them to be taken away in due course. The holders of the permits were not protected against them being taken away by government. What happened was that the permits privatized the public ranges. They created 'an odd species of property', 'less than a right but more than a mere revocable privilege.[2] Ranchers' political clout meant that the Bureau of Land Management 'acquiesced in the creation of de facto private rights in the public rangelands while neglecting to improve range condition.[3] Rather than hastening, tightening, streamlining and economising on environmental protection, the permits resulted in a different dynamic between regulators and regulated,[4] in which those to be regulated gained a whole new raft of rights to do as they wished with the land they had under permit.

In 1990 the US Clean Air Act Amendments, which launched the sulphur dioxide trading programme, was another example of temporary rights becoming permanent. The US government encountered an obstacle: the 'takings clause' of the Fifth Amendment of the US Constitution which prohibits 'private property' from being 'taken for public use without 'just compensation'. In order to avoid having to pay compensation to corporations for requiring them to reduce their emissions, the government faced a conundrum: Congress was concerned to defend corporate privilege in a working market and would not take kindly to the government trying to deny that emission permits were in fact property, as the law specified.

WHO HOLDS WHAT RIGHTS?

Under a scheme advocated by many economists, permits are sold to polluters by government. Under another scheme backed by many environmentalists, they are given to a trust which sells them to polluters at intervals and distributes the revenue to citizens. But under most real-world trading schemes, including US pollution trading programmes, the Kyoto Protocol and the EU Emissions Trading Scheme, they are given to a selection of historical polluters – wealthy countries and companies – for free.

---

2 Dennis, J.M., 'Smoke for Sale: Paradoxes and Problems of the Emissions Trading Program of the Clean Air Amendments of 1990', *UCLA Law Review* 40, 1993, p 1125
3 Ostrom, E., *Governing the Commons: The Evolution of Institutions for Collective Action*, Cambridge University Press, 1990
4 Ibid.

The US acid rain programme, for instance, handed out sulphur dioxide emissions rights free of charge to several hundred large industrial polluters – companies such as Illinois Power and Commonwealth Edison. The Kyoto Protocol dispensed greenhouse gas emissions rights to 38 industrialised countries who were polluting the most already. Although the South was allowed to continue emitting greenhouse gases unimpeded for the time being, it got no allowances to trade. The first phase of the European Union Emissions Trading Scheme (ETS), which got under way in 2005, donated carbon dioxide emissions rights to 11,428 industrial installations, mostly in the high-emitting private sector.

In other words, like rights to many other things that have become valuable – oil fields, mining concessions, the broadcast spectrum – rights to the earth's carbon-cycling capacity are gravitating into the hands of those who have the most power to appropriate them and the most financial interest in doing so.

The United Nations would never give away a public good to rich nations. European governments would never give away rights to the global carbon dump to its own corporations. Who would allow such a thing to happen?

It has already happened. The Kyoto Protocol gave Germany, France, Sweden and the rest of the European Union formal, transferable rights to emit, in 2012, 92% of what they were emitting in 1990. Japan and Canada got 94%, Russia 100%, Norway 101%, Iceland 110%. Under the EU's ETS, the UK government alone hands out free, transferable global carbon dump assets worth around £4 billion yearly (at June 2006 prices) to approximately a thousand installations responsible for around 46% of the country's emissions. Saleable rights to emit 145.3 million tonnes of carbon dioxide per year were given out to power generators, 23.3 million tonnes to iron and steel manufacturers, and so forth.

These emissions trading programmes are giving out 'allowances', not rights to pollute. The Marrakech Accords – the 'rule book' for the Kyoto Protocol – states clearly that the Protocol 'has not created or bestowed any right, title or entitlement to emissions of any kind on Parties included in Annex I'. The EU's ETS creates discrete permits under a regulation, not property rights. The US Clean Air Act Amendments of 1990 are likewise careful to specify that a sulphur dioxide allowance 'does not constitute a property right'. No one's giving anything away to polluters.

If only it were so! In fact, things are more complicated – and more disturbing. When governments say they are not giving out property rights, what they mean is that they are not giving out a particular kind of property rights.

What do you mean, 'property rights of another kind'?
Let's begin by acknowledging that there are good reasons why governments are afraid to mention the words 'property' and 'rights' in laws and treaties governing emissions trading.

An emissions trading system has to cut emissions and prove it is doing so. It can do that only if it reduces the amount of pollution allowances in circulation. Governments have to be able to confiscate some of the emissions allowances they gave out previously. And they have to be able to confiscate them without compensating their holders.

Why?
Imagine what would happen if the government had to compensate permit-holders every time it tightened an emissions 'cap' by taking away some of their allowances. Taxpayers would have to pick up the bill for every emissions reduction that corporations made, and the bill would be 'prohibitively high'.

In a housing market, homeowners need to know that the government can't simply take away from them their rights to their houses without compensation and sell the houses, pocketing the proceeds itself. But in an emissions market, it's essential that the government does have the power to take away some of the rights to pollute it has given or sold to companies or individuals. The property rights in an emissions market, in other words, must be less 'absolute' than the property rights in a car market. And in the case of carbon trading, it's especially important that governments be tough about taking away allowances.

Why will governments need to be tough about taking away carbon trading allowances?

Because they're going to have to take away so many in order to forestall climate chaos.

In the first phase of the Kyoto Protocol, governments have handed out, to industrialised countries alone, several times more rights to the world's carbon cycling capacity than would be available if global temperatures are not to rise by more than, say, two degrees centigrade. Having given a temporary stamp of approval to this huge overflow, governments will have to commit themselves to taking away an especially large proportion of those rights in the future.

Unfortunately, the rights holders in question – powerful Northern governments and their heavy industries – are not going to give them up without a fight. In fact, the fight has already started. So the job of dispossessing them of their carbon emissions permits not only carries much higher stakes, but will also be politically much harder for the UN and world governments to carry out, than the job the US government faced in taking away sulphur dioxide permits.*

---

*   Eds: Lohman, Hällström, Nordberg and Österbergh, 'Carbon Trading, a Critical Conversation on Climate Change, Privatisation and Power', *What Next*, issue No. 48, 2006, pp. 73 – 77

Chapter 6

WHEN *MALUM IN SE* BECOMES *MALUM PROHIBITA*

"Change begins with a sense of stewardship which itself grows out of understanding. The more we seek to know our environment and the history of our relationship with it, the more we will see a much clearer course for change. The concern that grows out of understanding should fuel our desire for concerted action. The time has come for us to make hard and judicious choices that will have a great effect on those who come after us. The care and responsibility we show for our water demonstrates much about our values, including our level of concern for the quality of life for future generations."

GRANT MACEWAN, *Watershed: Reflections on Water.*

MONEY is like water. It can nourish life; it can flow into life-affirming ventures, providing the very nourishment required to help them grow and flourish. Or it can be used to sustain damaging projects, projects that create toxic and polluting environments.

The Oil Industry

There are two terms in law that we rarely hear today: *malum in se*, which means 'wrong in itself' and *malum prohibita*, which means 'wrong because it is prohibited'. There is a crucial difference between the two – the first is a moral premise; the second is a legal one, often based on a moral premise that has been adopted as law. When our laws are built from *malum in se*, we have the makings of a higher moral code.

The Canadian Tar Sands can be termed *malum in se* and given time will become *malum prohibita*. According to the International Energy Agency, the Canadian tar sands production would be curtailed if policies to address climate change were implemented.[1] Two salient factors would come into play: 1) high carbon emissions from tar sands production would become untenable; 2) tar sands production growth would become uneconomic in a low carbon economy. Named the 'marginal barrel' on the oil market, tar sands production will lose to any concerted efficiency drive. Analysts have expressed concern about the 'narrowing window of profitability' in unconventional oil extraction, citing increasing constraints on resources, such as water and natural gas, as adding to the already high costs, and future environmental regulations.[2] Costs related to land reclamation have been underestimated and have only been nominally addressed, increasing the likelihood of large expenditure in the future.[3]

All the major oil companies have involvement and investment in a range of tar sands reserves and projects. Yet the vulnerability of tar sands production growth was demonstrated in 2008 when following the oil price crash to $35 per barrel in the last quarter of that year, all projects that had not yet broken ground were put on hold. Of two million barrels per day (mb/d) of non-OPEC production capacity that was shelved in this period, 1.7mb/d, or 85%, was Canadian tar sands production.[4]

---

1   *World Energy Outlook*, International Energy Agency, 2010 report
2   *Canada's Oil Sands: Shrinking window of opportunity*, CERES Risk Metrics Group, report May 2010.
3   *Toxic Legacy: How Albertans could end up paying for oil sands mine reclamation*, The Pembina Institute, September 2010.
4   *Medium Term Oil Market Report*, IEA (2009)

The tar sands industry is unviable when oil prices fall below $60 a barrel and possibly needs prices over $100 a barrel to realise the growth desired.[5] While Canadian tar sands present little exploration risk, producing oil from the viscous bitumen is currently among the most expensive commercial oil production processes. Construction of a typical tar sands project requires enormous capital expenditure. The latest tar sands mine to come on stream, Shell's Jackpine mine, cost $14 billion (bn) to create 100,000 barrels per day (b/d) capacity and was cited by Shell's head of tar sands production as being 'some of the most expensive production that we have.'[6]

It will require a minimum oil price of $70–75 a barrel to turn a profit. High operational costs and low returns, especially if not integrated with upgraders, spell disappointing results and, for many *in situ* projects, low recovery rates plus a poor operating cost factor of steam-to-oil ratio.[7]

Lack of water leads to violence; it is a fundamental and basic resource which when limited can lead to conflict.[8] Darfur was a war over water and many nations that are now suffering water poverty for the first time in decades are finding it the source of many problems. Without water, all life begins to suffer – both the food that needs to grow and the people and the land upon which they depend.

Deepwater

The tragedy that unfolded through the summer of 2010 for BP and the Deepwater Horizon served to highlight in a very public arena

---

5 *Oil sands industry update: production outlook and supply costs 2009–2043*, CERES, May 2010 and Canadian Energy Research Institute, November 2009. Prices are in 2009 dollars.

6 'Shell puts oil sands expansion plans on hold'. *The Globe & Mail*, (28 April 2010) *In situ* production involves injecting steam into the bitumen reservoir to fluidise the bitumen so it can flow through a production well. The ratio of steam needed to produce a barrel of oil is key to the economics of *in situ* production as a high steam requirement raises the cost of inputs, primarily natural gas.

7 Roche, P., 'SAGD: Spin vs Reality; Eighty-Two% Of Alberta's Oilsands Output Must Come From Wells, And The In-Situ Technology Of Choice Is SAGD. How Good Is It?', *New Technology Magazine* (1 September 2007)

8 See *World Water Development Report* and Ban Ki-moon's statement on World Water Day 2010: 'More people die from unsafe water than from all forms of violence, including war". http://www.un.org/apps/news/story.asp?NewsID=34150&Cr=water&Cr1

the high risk involved in drilling in such an extreme environment. As a result the cost of ultra-deepwater oil production has now risen.

Prior to the disaster, the push into ever deeper water and more complex operations was raising costs and risks to the point that something had to give. Blowout prevention procedures had been identified as inadequate by the now defunct Minerals Management Service and the deepwater industry in the Gulf of Mexico. Indeed, the difficulty of controlling a blowout in the ultra-deep had been raised on a number of occasions over six years.[9] As one ex-industry executive commented: 'Our ability to manage risks hasn't caught up with our ability to explore and produce in deep water'.[10]

While the complexity of operations grows with depth and remoteness from shore, so does the difficulty of addressing a leak. There is little evidence that anything less than a relief well can guarantee closure of a blowout at these depths. Whilst the Deepwater Horizon Macondo well sits at a depth of 4,993ft (1,522m), it is not the deepest offshore production: Shell operate the new Perdido platform in water depths of 8,000ft (2,450m). The pre-salt deposits off the Brazilian coast present further challenges as they are not only in ultra-deepwater but are also thousands of metres below the seabed under layers of rock and unstable salt.

BP, Shell, Chevron and ExxonMobil have all invested enormous sums in deepwater production. Indeed deepwater, and more recently ultra-deepwater, production has been the primary target for much of the industry in the 21st Century. According to IHS CERA, 'from 2006 to 2009 annual world deepwater discoveries (600 feet or more) accounted for 42–54% of all discoveries – onshore and offshore. In 2008 alone, deepwater discoveries added 13.7 billion barrels of oil equivalent to global reserves'.[11]

Deutsche Bank issued a research note soon after the Horizon spill which highlighted the importance of deepwater production

9   Campbell, R., *Special Report: Deepwater spills and short attention spans*, Reuters, (14 June 2010) www.reuters.com/article/idUSTRE65D3Z220100614

10 Chow, E.C., quoted in Mouawad and Meier, 'Risk-Taking Rises as Oil Rigs in Gulf Drill Deeper', *New York Times* (29 August 2010)

11 *The Role of Deepwater Production in Global Oil Supply*, IHS CERA (30 June 2010)

to the industry and speculated on the impacts of the disaster. The analysts noted that 'The Gulf of Mexico was the biggest growth-driving region of the biggest growth-driving oil supply theme: the deepwater'.[12] Commenting on the importance of deepwater, they noted that it 'has been the source of over 60 billion barrels of 2P (proven plus probable) reserves since the late 1990s and [...] today represents the leading single source of growth in oil production with anticipated volume growth of some 12% p.a. over the 2000–2015 period against just 2% for the market overall, directly as a result of the technological and physical push to deeper and deeper water, previously inaccessible to drilling and production'.[13]

The longevity of some deepwater fields has come under scrutiny after recent experience from some wells cast doubts on future use. BP's Thunder Horse project in the Gulf of Mexico is a case in point. Having been delayed by technical and engineering difficulties for four years this ultra-deepwater project started production in June 2008 following $5 billion of investment. Its capacity was projected to be 250,000b/d but that level has never been reached and it hit its production peak within six months of coming on-stream.[14] By February 2010, water was making up 31% of the liquids being produced at the main well.[15] It is doubted that it will reach its lifetime target production of one billion barrels.[16] Similar problems were experienced at BHP Billiton's Neptune project and a further twenty-five deepwater wells have similarly experienced rapid decline.[17]

THE HIDDEN WORLD OF SILENT RIGHTS

As legal scholar David M. Driesen of Syracuse University's School of Law puts it, there is a 'tradeoff' between the 'need to

---

12 *Macondo and the Global Deepwater*, Deutsche Bank Global Markets Research (June 2010)
13 Ibid p.18.
14 Morton, G., 'BP's Thunder Horse to Under-Perform in the Wake of the Deepwater Horizon Blowout?' *The Oil Drum*, (30 April 2010) www.theoildrum.com/node/6415
15 'The Other Deepwater Oil Production Problem:The Decline of Thunder Horse Field', *Energy Facts Weekly* (17 May 2010) www.energy-facts.org/LinkClick.aspx?fileticket=Jz0MG1DqfME%3D&tab id=100
16 Morton, G., Ibid
17 Steffy, L., 'In deep, and falling far short: Deep-water output worrisome', *Houston Chronicle* (16 May 2010)

protect the public properly from environmental harms that may grow over time' and 'stability to encourage cost-decreasing trades.'

For the market to work at all, 'interests in allowances must be sufficiently protected to protect investment.' Indeed, guaranteeing that 'property rights can be assigned and enforced to ensure that trades can take place in an ordered fashion and with a high degree of certainty' is the 'key role of the policy system' in an emissions trading scheme. Nobody who holds emissions allowances, or is thinking of buying or selling them – whether polluter, broker, banker or investor – is going to want anybody to be able to take them away arbitrarily.

So just as corporations lobby for exemption from pollution regulations, they lobby to make sure emissions allowances amount to secure property rights and to get as many as they can. As 'semi-permanent property rights,' in the words of David Victor of the US Council on Foreign Relations, emissions permits are 'assets that, like other property rights, owners will fight to protect.'

Luckily for corporations, their privileged access to legislators enables them to secure carbon dump permits for themselves merely by lobbying and pressure politics. Just as systems of private property in land give new moneymaking powers to surveyors, officials and firms with access to titling and licensing mechanisms, the property systems of pollution trading schemes give new commercial powers to those with access to legislators.

As economists Peter Cramton and Suzi Kerr point out, the 'enormous rents' at stake 'mean that interest groups will continue to seek changes in the allocation over time':

> Firms may end up putting as much effort into rent capture as into finding efficient ways to reduce carbon usage. Investments may be delayed in the hope that high observed marginal costs would lead to more generous allowance allocations as compensation. The increased complexity of the programme... may lead some groups to seek exemptions or bonus allowances... [I]nterest groups will fight bitterly for a share of annual rents. This fight will lead

to direct costs during the design of the policy. Groups will invest in lawyers, government lobbying, and public relations campaigns. Government officials will spend enormous amounts of time preparing and analysing options and in negotiations. This will lead to high administrative costs and probably considerable delays in implementation.

Governments eager to placate industry are almost sure to give out too many emissions rights. This in turn will make future cuts even more difficult, while increasing pressures to reduce emissions in sectors that have not been awarded rights (for example, domestic households, the transport sector and the state).*

---

* Eds: Lohman, Hällström, Nordberg & Österbergh, 'Carbon Trading, a Critical Conversation on Climate Change, Privatisation and Power', *What Next*, issue No. 48 (2006) pp. 81, 82

....

To clean up the Deepwater Horizon spill, BP sprayed approximately 771,000 gallons of a toxic chemical dispersant called Corexit 9500A. A study, which was published in *Environmental Science & Technology*, found that, contrary to popular belief, the dispersant did not degrade but instead moved with the oil plumes until at least September 2010. Deepwater applications of dispersant have not been tested and thus no data exists on the environmental impact of dispersants such as Corexit 9500A on the sea. When the Deepwater Horizon platform exploded and sank in the Gulf of Mexico in April 2010, its uncapped deep sea well fed millions of gallons of oil into the ocean. Specific research of Corexit performance does not exist so the outcome is yet to be determined, despite claims that chemical dispersants would help break down crude oil and eventually degrade in the water. In the event, the dispersant was deployed liberally. The oil remained on the surface for months after the well was capped, drifting instead of biodegrading.[18]

---

18 http://www.care2.com/causes/environment/blog/bp-oil-spill-dispersants-drifted-but-didnt-degrade/

DOCTRINE OF SUPERIOR RESPONSIBILITY

'That military commanders and other persons occupying positions of superior authority may be held criminally responsible for the unlawful conduct of their subordinates is a well-established norm of customary and conventional international law. This criminal liability may arise either out of the positive acts of the superior (sometimes referred to as "direct" command responsibility) or from his culpable omissions ("indirect" command responsibility or command responsibility *strictu sensu*). Thus, a superior may be held criminally responsible not only for ordering, instigating or planning criminal acts carried out by his subordinates, but also for failing to take measures to prevent or repress the unlawful conduct of his subordinates.

'The distinct legal character of the two types of superior responsibility must be noted. While the criminal liability of a superior for positive acts follows from general principles of accomplice liability the criminal responsibility of superiors for failing to take measures to prevent or repress the unlawful conduct of their subordinates is best understood when seen against the principle that criminal responsibility for omissions is incurred only where there exists a legal obligation to act. As is most clearly evidenced in the case of military commanders by article 87 of Additional Protocol I, international law imposes an affirmative duty on superiors to prevent persons under their control from committing violations of international humanitarian law, and it is ultimately this duty that provides the basis for, and defines the contours of, the imputed criminal responsibility'.*

---

* Prosecutor v. Delalić et al. (Čelebići case), Judgement, Case No. IT-96-21-T, T. Ch. IIqtr, 16 November 1998, paras 333 and 334

....

Superior responsibility is a principle well known to lawyers of international law, in particular the law of war. However, it is a principle that has been applied to others not involved in war. Most notably, in the trials set up in Nuremberg, directors of companies that profiteered out of the war, such as Friedrich Flick, were prosecuted for taking decisions that had direct adverse impact on those whose right to life was being breached. Six members of the Flick Concern, a group of industrial enterprises (including coal mines and steel plants) were charged and prosecuted for international crimes. Each of them was in a position of superior decision-making within their companies and were therefore responsible for the actions taken by others under their jurisdiction.

The Ecocide Act, as set out in the Appendix II of this book, has been drafted to include the principle of superior responsibility for all who are in a position of authority, and who are in a role which imputes a mantle of responsibility that takes precedence over others within the company structure. Those few persons who sit in the seats of control are the ones to whom we look for ultimate decision-making. Their roles carry the explicit duty to ensure no harm comes not only to those who work for them, but also to others whose lives can be adversely impacted by the resolve of a few who have power to change the lives of many across the world.

We shy away from the word 'value', even though all that we do is driven by our values, whether they be constructive or destructive. Every decision we make in life is informed by our values. From them come our decision-making which in turn, on a political level, drives our policies and practices that govern society for many years. Sometimes we need to challenge our values when it becomes abundantly clear that they are not serving us well. A well run business will have in place workers who will be able both to look ahead and to pre-empt any potential explosion: a business which cuts costs to bolster their primary value – profit – has stepped into the realms of neglect.

Neglect is a state of being where wider considerations have been forfeited in favour of a more immediate concern. Often the concept of neglect is applied to how we look after our home, our family and friends, but it applies equally in the macro world of business. Just as neglect of childcare raises welfare issues, so does neglect within a workplace. Just as neglect of the land you live on leads to an untended garden that soon becomes unmanageable, so can a lack of attention to the land you are working on damage the environment. What happens in the backyards of our homes mirrors the backyards of our workplaces. It is no coincidence that 'economy' means 'management of our home'. When we fail to manage our household affairs, our lives and those of our dependants become adversely affected. Neglect can become a way of being which soon becomes the norm. It is a term which applies in family law: where a parent has neglected their child, the state can step in to remedy the situation

The Deepwater Horizon oil spill was an obvious example of a tragedy born of a history of decisions to cut costs to ensure a lean operation unburdened by too many safety restrictions. Various environmental regulations do not add up to a law that prohibits ecocide: BP were free to destroy a vast tract of sea and its inhabitants without having to pay the true price for the damage to many species. What was valued was protection of their business, even where that caused yet more damage to others. Intrinsic values, such as caring for the sea to ensure that no further damage was done, was not a primary consideration. Intrinsic values, such as caring for the community of fishermen adversely affected by the pollution, was not a primary consideration. Intrinsic values, such as the well-being of the fish, was not a primary consideration. Extrinsic values dictated BP's response: limiting the cost of clean-up was the primary concern. Use of chemicals to sink the oil below the waterline was an economic decision not an environmental consideration. Had it been, another way could have been found to remediate the sea. As it was, no remediation has been put in place.

Businesses that fail to take responsibility for the risks they carry often do not have systems in place to ensure damage and destruction is restricted. In pursuit of short-term profit, driven by a value that would never be applied to our home-life, we sacrifice much of life around us. Apply profit-margins to your family and soon the cracks appear. Whilst good household management requires sound budgeting, decisions are made first and foremost on the values of care, health and well-being. Place these values at the centre of business and we have the recipe for long-term success.

When we balance the needs of civilisation against the needs of the profit and loss account and look to the future, it is not difficult to see that short-term gain soon disappears when it is measured against the long-term outcomes for the earth. Profit that is predicated on destroying the earth to make money cannot be a benefit in the long-term. Profit that is predicated on caring for the earth benefits all. Care for the earth, or at the very least the patch of the planet that your business is working on, and you will find that when matters go wrong, you know what to do to ensure the best outcome for all.

The Deepwater Horizon oil disaster arose from a legacy of cuts and measures to optimise profit margins by outsourcing when BP did not have the necessary in-house expertise. They were therefore at a loss as to how to deal with the catastrophe that hit them. Instead of taking care, the buck was passed onto others. However, in legal terms, the duty of care is not forfeited entirely by simply giving the baby to another: one has to ensure the surrogate parents are fit for purpose.

VOLUNTARY REGULATION OF SUBSTANCES THAT CAUSE ECOCIDE

Another ecocide is playing out in neighbouring Canada; the Canadian Tar Sands are possibly the world's largest ever ecocide and yet they remain largely hidden from public view. The Standing Committee on Natural Resources had recommended in 2007 the acceleration of treatment of toxic waste water from the extraction of oil from the Athabasca tar sands and recommended that industry should take the lead in speeding up research and action to reclaim land rendered toxic by the tailings and water. Despite the size, duration and impact, it has been allowed to continue. Instead of outlawing it, the government proposed two soft solutions: voluntary action and further research by the industry polluters. The Alberta Energy Resources Conservation Board has now directed that the tailings be reclaimed by specified dates, However, they have extensive powers to grant exemptions which will render the process toothless. A litany of concerns about the failure of federal authorities to implement and police effective regulations had been presented to the Committee. Despite the level of concerns, no response was given to complaints about harmful air emissions and effluence emitted and released by the oil sands sector. The reality is that lax regulations have paved the way for rapid expansion of the sector with sizeable profits to be gained. A number of federal laws, such as the Canadian Environmental Protection Act (CEPA) and the federal Fisheries Act, to regulate harmful toxins have been ignored and no action has been taken.[19]

Published studies identify that significant volumes of contaminants emitted by the unconventional tar sands sector are entering the environment.[20] Yet few legally binding standards have been imposed and curtailment of significant carcinogens and heavy metals emitted by the sector which are reported to be accumulating in the watershed have been ignored. Despite the evidence of damage and destruction as well as contamination, the setting of standards have not been implemented in the oil sands sector.

The oil sands sector produces and emits significant volumes of contaminants via its operations. The process of separating oil from the sands produces contaminated waste water and tailings, licensed to be stored in tailings ponds. The National Pollutants Release Information (NPRI) database reveals that 10% of mine tailings in Canada are produced by oil sands. This is all the more critical given the toxicity of the liquid tailings stored in ponds covering about 200 square kilometers and containing oil and grease, naphthenic acids, cyanide, arsenic, hydrogen sulphide, phenols and a litany of measurable heavy metals, including mercury, chromium, vanadium and cadmium.

An estimated 720 million cubic meters of impounded toxic liquid tailings are sprawling across the landscape, covering over 130 square kilometers of land that used to be ancient peat land. It is now filled with toxic waste generated from the tar sand extraction. It is death by a thousand cuts. Each toxic tailings pond brings

19 *Missing in Action. New Democrat Report on the Standing Committee Review of the Impacts of Oil Sands Developments on Water Resources; Key findings: Regulation of Toxic and Deleterious Substances*, p.13-14

20 Erin N. Kelly, et al, 'Oil sands development contributes polycyclic aromatic compounds to the Athabasca River and its tributaries', pnas.org/cgi/doi/10.1073/pnas.09.0912050106; Timoney, Kevin and Lee, Peter, 'Does the Tar Sands Industry Pollute? The Scientific Evidence', *The Open Conservation Biology Journal*, 2009, 3, 65-81.

with it risks of discharge, seepage and long term impact on the surrounding fragile ecosystems. A preliminary summary based on 2006–2009 data reports arsenic and lead levels to have increased by 26%. Complete data is yet to be revealed by the sector.

Scientific studies confirm that air emissions from bitumen processing are impacting the watershed, the fish, their food-chain and it threatens the larger Mackenzie River Basin. While industry testifies that there is no unrecovered seepage, others have suggested a conservative estimate of tailings seepage is in the order of eleven million litres per day.[21]

We have two routes: voluntary initiatives or mandatory laws. Within those two options, we have several alternatives already proposed. Voluntary schemes include trading mechanisms, such as Reducing Emissions from Deforestation and Forest Degradation (REDD). Another option is voluntary contributions, such as the Yasuni Ishpingo Tambococha Tiputini (Yasuni ITT). This is a proposal by the government of Ecuador to refrain indefinitely from exploiting the oil reserves of the Ishpingo-Tambococha-Tiputini (ITT) oil field within the Yasuni National Park, in exchange for 50% of the value of the reserves. However the mandatory route has, unlike voluntary agreements, the force of law. Mandatory law halts the damaging and destructive activity upstream by leveraging sanctions. Sanctions, depending on their severity, can be the most effective mechanism of all.

Voluntary trading mechanisms such as REDD places ownership and control of forests with those who stand to gain financially. As we have seen, corporate entities who obtain the property rights to the trading of the forest emissions can determine their fate. When national governments cede their natural heritage to the commercial interests of business, REDD+ does nothing to protect the interests of those who live there. In so doing, control

---

21 Testimony from Dr. David Schindler and Dr Kevin Timoney, Acting ADM Environment, Canada, March 12, 2009.

of the rainforests and the plantations will be determined by those whose primary purpose will be financial gain from trading their carbon commodity. This option allows corporate ecocide to continue unabated. Voluntary offsets, such as REDD+, create a framework within which global north nations continue to collude in the destruction of territory elsewhere.

Voluntary contribution measures, whilst seeking to protect the rights of the indigenous communities, are non-enforceable and unlikely to be given anything more than the semblance of consideration. Any verbal agreement will be vulnerable to the retraction of those who have pledged assistance.

Mandatory emission caps, is premised on the imposition of fines upon those who fail to comply. Recent evidence has however demonstrated that fines have done nothing to abate global limits, whether it be to abate national greenhouse gas emissions or to limit the rate of logging in the Amazon. The Emissions Gap Report[22] tells us what will happen if we follow the existing path: the world will overspend its remaining carbon budget by a third by 2020. Whilst The Emissions Gap Report emphasises that tackling climate change is still manageable, but depends upon financing markets to address mitigation and adaptation, no action has yet been take to prevent pollutants at source.

The fourth option is to criminalise ecocide. The destruction of vast tracts of land by corporate activity has been normalised by contracts that assign silent rights to businesses. In so doing, the rights of the wider Earth community are sidelined. Without realigning the balance of interests, corporate ecocide will continue unabated and very soon we will no longer be able to sustain our communities ecologically. Many communities are at risk of losing their ways of living and life itself where governments remain silent rather than join the growing support to change the agenda and push for laws to protect, prevent and restore. By criminalising ecocide, the defunct UN Trusteeship Council can be reopened,

---

22 Levin, K. and Ward, M., *The Emissions Gap Report: Are the Copenhagen Accord Pledges Sufficient to Limit Global Warming to 2 degrees Celsius or 1.5 degrees Celsius?* United Nations Environment Program (2010).

this time to impose a legal duty of care on all nations to assist those territories at risk of ecocide. By way of example, by applying under the UN Trusteeship Council to make the Yasuni initiative in Ecuador a UN Trust Territory, all nations could be legally bound to assist.

Global problems, such as climate change, transnational pollution, competing nations, unequal food distribution and banking without ethics, all stem from property laws. Rationality prevents all semblance of restraint and cooperation. The shift in business motivation we are now tasked with accomplishing is a far deeper and more encompassing shift than just closing the door to one practice. Making sure that one door is closed allows another to open. This is about stepping through to a new paradigm and changing the values that drive our businesses. When we change the values a different way of doing things inevitably emerges; our priorities shift to a world view that places the interests of peace first.

Where health and well-being of people and planet are achieved, the path to peace can emerge. Business has a role in creating world peace through its leadership and best practice. A company that adopts a process of evaluating whether their work is in the interests of all, rather than serving only a few, has an opportunity to find new solutions that ensure long term benefit to the wider community, not just their shareholders pockets. Rather than fostering conflict, peace can prevail.

Peace is an elusive concept: setting the conditions for it to flourish requires the good health of business, the people who work there, those who are affected by the business – whether they be the local community or future generations – and the nature of the business itself. A cognitive dissonance arises where people and planet are being treated differently. By keeping the same values that apply to our human relationships and extending our circle of concern outwards, we can embed a holistic approach to health and well-being, a deeper systemic relationship between all parties.

Part 3

TOWARDS A NEW WORLD

*Only when we face the shadow self and
give it a name, does the healing begin*

Chapter 7

TRANSITIONING INTO A NEW ERA

Valuing Nature

'Are there other socially viable paths for conservationists besides the commodification of nature? Yes. Nature has an intrinsic value that makes it priceless, and this is reason enough to protect it. The idea is not new. We view certain historical artefacts and pieces of art as priceless. Nature embodies the same kind of values we cherish in these man-made media. Some ecologists claim that these intrinsic values, often referred to as cultural services, figure prominently enough in their valuation programmes.'[1]

Valuing nature for its own sake is not an idea that fits easily with a system that has placed profit as its primary goal. The temptation to put a price on it so that we can show it in a graph is all too real, and now the valuing of nature has become a means to quantify the unquantifiable. It is so easy to relegate nature to a number in a column of figures where it can be added or subtracted according to the ecological credit or debit we have accrued. It is also so easy to hide the true cost.

---

1   Terborgh, J. *Requiem for Nature*, (1999)

Politicians are talking about ecosystems services and so are the media. The civil rights movement in the 60s refused to use monetisation of blacks as a means to modify discrimination. Martin Luther King and others refused to engage in that debate: they reasoned that it stopped recognition of the intrinsic value of humans regardless of race, colour or creed. The same can be said for ecosystems: it stops the enactment of laws to protect nature based on its intrinsic value of life for all. Out of the civil rights movement came abolition of apartheid, anti-discrimination laws and criminal legislation to prohibit racist crimes. None of this happened on the basis of monetising the civil rights movement. There was an inherent understanding that to commoditise the problem would ultimately lead to greater polarisation in society. To argue they were 'worth x' to the economy was in effect to create a new slavery, where blacks became yet again a commodity with a price on their heads.

Students of economics who are today taught that all markets work by monetisation, are likely to have stronger extrinsic values than, for instance, biology students. However, the indicators of a changing world are already here: universities, such the University of Uppsala in Sweden, and business schools, such as the one in Exeter, England, are changing the paradigm student by student and creating a new generation of business leaders who are driven by intrinsic values, not by monetising their decisions. Creating a green economy is not merely about putting subsidies in place; it is about identifying the debt that needs to be repaid and facing the new paradigm squarely and embracing it.

Education has a crucial role in tooling-up a new generation who will have a set of expectations and a framework which differs from the current norm. For them, it is *their* norm – they engage and undertake business placements with business leaders and mentors who are already changing the way business is run. Their sense of purpose is aligned with their values and very quickly it becomes apparent that there can be other ways of doing business – without destroying the earth. The process of education in these forward-thinking establishments becomes intergenerational, whereby the student teaches the teacher and inputs into the course to suggest improvements that could benefit the next generation of students.

By giving students their own space to self-determine what the course needs to cover brings with it two very vital tools; a sense of giving back to the system and a wider circle of concern, looking both to the future and the past. Stewardship of the course engenders a sense of responsibility to those yet to become part of the course. By fostering a relationship with course-providers, students have a role in ensuring the lasting success of their programme. They become stewards for future generations. Not only do they learn the skills of best business practice, they also acquire the habit of giving back to the system that gave to them. These are the skills which make the leaders of the future who lead by example and in service to the wider community. Those who participate have a far higher success rate in their chosen field and often go on to become change-makers in their own right.

Rapid change is effected in various ways. The embedding of values which are focused on wider concerns can change the workplace very quickly. Soon students will be looking over the horizon to the world they want to see. Their vision will have been expanded and they will not brook a system over which they have no control. Instead of assimilating the existing system, they will co-create the world they want. One very potent example of the rapidity with which students and young people adapt when confronted with a new world paradigm was when the Berlin Wall came down. The impact on social institutions in East Germany was dramatic: they converged rapidly to western values. The political context and institutions reinforced it and elicited responses which prevented the old systems from retaining their iron grip.[2]

Results show a considerable convergence in attitudes between Eastern and Western Germany – attitudes in Western Germany remain stable while attitudes in Eastern Germany converged, over time, to those found in the West. Comparisons of those Germans with birth dates pre- and post 1989 show that, while considerable

---

2 Svallfors, S., 'Policy feedback, generational replacement, and attitudes to state intervention: Eastern and Western Germany 1990-2006', *European Political Science Review*, 2, (2010) pp.119-135

attitude differences between Eastern and Western Germany were still found in 2006 among those who had their formative experiences before the fall of the Wall, differences are virtually nil among those who were still children in 1989. Analysis of this period provides evidence for the attitude-forming effects of events which act as catalysts for change. It also points to generational replacement as a key mechanism in translating institutional change into attitudinal change.

A bastion of light in the conservation world is to be found in Central America. Costa Rica has developed what Terborgh refers to as a 'new paradigm' for tropical conservation.[3] Unfortunately, by the time Costa Rica adopted its conservation model, largely based on the US strategy of people-free parks, only fragments of forest remained after vast tracts had been destroyed by human activity.

One of the highest priorities is the Amazon (a forest region half the size of continental United States) of which over 15% is already lost through commercialisation. The Amazon and other large native forests are crucial to ensure the seventy-five billion tons of carbon which are stored in the forest biomass remain out of the atmosphere and in the earth. Seven trillion tons of water from the forest evaporate into the atmosphere each year, forming rain clouds which feed the bio-diverse region. Once the forests go, the climate of South America will change in a matter of decades, converting the continent from its verdant, luxuriously-vegetated state, the legacy of millions of years of extraordinary evolution, into one that resembles Africa, with its deserts, semi-arid areas and savannahs. Where large expanses of forest cannot perform these functions, the rest of the world cannot function. To save them from extinction must surely be one of our top priorities: to lose them would be an enormous ecocide.

THE FORESTS OF GUANACASTE

In the time it takes you to read this page, some thirty-two hectares of the world's tropical rainforests will be destroyed – a statistic that

3   Terborgh, J. *Requiem for Nature*, Island Press, Washington DC, 1999.

defies comprehension. One hundred years ago, rainforests covered two billion hectares, 14% of the earth's land surface. Now only half remains, and the rate of destruction is increasing: an area larger than the state of Florida is lost every year. If the destruction continues apace, the world's rainforests will vanish within forty years.

Costa Rica's conservation programme has the potential to spread right across Central America, away from the soil-leaching deforestation that plagues the isthmus. The country has one of the world's best conservation records: about one-quarter of the country is under some form of official protection. In 1992, Costa Rica received the Cantico a Todas Las Criaturas (Song to all Creatures) award given by the Franciscan Center for Environmental Studies, based in Rome. In April 1992, the National Biodiversity Institute was also awarded the Peter Scott Award by the International Union for the Conservation of Nature. Despite Costa Rica's achievements in conservation, almost the entire country has been deforested and deforestation continues at an alarming rate.

The Cost: While many patches of forest will no doubt be saved, many animal and plant species can only survive in large areas of wilderness. Most of the millions of rainforest species are so highly specialized that they are quickly driven to extinction by the disturbance of their forest homes. Isolation of patches of forest is followed by an exponential decline in species. The reduction of original habitat to one-tenth of its original area means an eventual loss of half its species. Eventually these biological islands become pauperised communities.

At the current rate of world deforestation, plant and animal species may well be disappearing at the rate of 50,000 a year. By the end of the 20th century, an estimated one million species had vanished without ever having been identified. Among them will be many species whose chemical compounds might hold the secrets to cures for a host of debilitating and deadly diseases. The bark of the cinchona tree, for example, has long been the prime source of quinine, an important antimalarial drug. Curare, the vine extract used by South American Indians to poison their arrows and darts, is used as a muscle relaxant in modern surgery. And scientists recently discovered a peptide secreted by an Amazonian frog called Phyllomedusa bicolor which may lead to medicines for strokes, seizure, depression, and Alzheimer's disease. In fact, some 40% of all drugs manufactured in the United States are to some degree dependent on natural sources: more than 2,000 tropical rainforest plants have been identified as having some potential to combat cancer.

Nonrenewable resources: Once the rainforests have been felled, they are gone forever. Despite the rainforests' abundant fecundity, the soils on which they grow are generally very poor, thin, and acidic. When humanity cuts the forest down, the organic-poor soils are exposed to the elements and are rapidly washed away by the intense rains, and the ground is baked by the blazing sun to leave an infertile wasteland. At lower elevations, humans find their natural water sources diminishing and floods increasing owing to removal of the protective cover given by the intact montane rainforest which naturally acts as a giant sponge. Thus, indigenous groups such as the Bribrí and Cábecar Indians who inhabit remote regions close to the Panamanian border are finding their tenuous traditional livelihoods threatened.

Conservation: Part of the government's answer to deforestation has been to promote reforestation, mostly through a series of tax breaks, which have led to a series of tree farms predominantly planted with non-native species such as teak. The government, for example, has extended legal residency status to anyone participating in reforestation programs, with a required minimum non-taxable investment of US$50,000. These efforts, however, do little to replace the precious native hardwoods or to restore the complex natural ecosystems, which would take generations to re-establish. Such efforts are being taken up by a handful of dedicated individuals and organizations determined to preserve and even replenish core habitats, such as attempts spearheaded by Daniel Janzen and the Friends of Lomas Barbudal to re-establish the tropical dry forests of Guanacaste.*

---

*   http://philip.greenspun.com/cr/moon/conservation

Community Resilience

A typical example of market-led planning is to be found on the outskirts of towns which have enormous infrastructure problems for anyone with a penchant for walking to the shops, work and their local pub. Luxury eco-apartments are feted for their green credentials, however their sustainability record is severely marred by the fact that a car is a necessary prerequisite to living in their comfortable and non-toxic home. It is ironic that, in fact, their carbon footprint will be exponentially increased due to their need to drive everywhere. They also create a social imbalance because only those who can afford them have access to them. The social fabric of society breaks down where the community becomes ever more fractured. To place people in housing developments, no matter how ecological they are, without basic amenities nearby produces isolation and disintegration. Those who are living on more basic means are often more resilient than those with more economic cushioning: in lower income neighbourhoods, by dint of close proximity, people have a greater sense of their neighbours than those who are surrounded by high brick walls and gated communities. Children at public and private schools often have no means of interacting with local children because of complex travel and school arrangements. Their parents, who have paid enormous sums for the right education, do not recognise the disconnect that such schooling can sometimes engender. Schoolchildren who travel far from their homes have a less keen sense of their immediate surroundings and have less of a network embedded with other children in the neighbourhood. Their social circle can often, like their parents, be very restricted by their travel arrangements. When problems arise, the basic fabric of the immediate community is not there.

Banff and Buchan College

Shell is working with Banff and Buchan College in Aberdeenshire as a sponsor for a unique technicians' training programme based on an HNC in Mechatronics. It was set up to attract young people, which has become increasingly difficult in recent years, into the oil industry. As a result, the industry has experienced a skills shortage when their experienced staff retire. David Cook, Technology Sector Manager, heads up the programme at the College. One year full-time training results in a HNC qualification, two years and it becomes an HND engineering course which combines electronic, mechanical and computer technology with first-hand experience. The course aims to develop knowledge of mechatronics, and thereby skilling up a new generation with access to a wide range of engineering opportunities in the oil industry.

Shell's contribution and knowledge of what employers want is vital to the development and improvement of this course. They have also built a testing platform at the back, to replicate what it would be like to work out at sea. Situated in Peterhead where there is lots of rain and wind, the students can get used to wearing thermal overalls and dealing with adverse weather conditions. What the Mechatronics course offers is first-hand experience that other courses don't provide. Shell are looking to the future and have seen that there is a skills shortage looming, so they are proactively addressing it at source. Students leave their course and go straight into jobs in an industry that has trained new employees with the skills needed to begin immediately. A huge incentive for these students is the opportunity to get some hands-on work experience. During the two-year course, students complete a five week placement with other companies. This valuable industry exposure gives them an opportunity of putting what they have learned into practice. Each student undertakes a different placement, where they are allocated a specific mentor and their progress is closely monitored. It's a very hands-on training: participation is key from the very outset.

When society fractures and begins to breakdown, the ability to share is compromised in the face of fear and mounting challenges. Without strong anchors within communities, resilience is difficult to embed in the communal fabric, especially after problems arise. Like building a house, communities need virtual foundations upon which to build a basis for the future without fossil fuel.

Communities that are facing their own local future, and building transitioning strategies are ahead of the game. They are fostering resilience and building skill-bases to ensure that help is at hand. By engaging in the issues before they arise, the networks begin to take shape within the community and the training begins from a small but powerful foundation. These communities are wanting to stop using polluting practices. They accept that environmental imbalance is largely the result of human-created pollution, including the generation of excess greenhouse gases. Greenhouse gases *per se* are not harmful – indeed our very survival is dependent on them – it is the imbalance that has detrimental consequences. However, a community taking action only gets them so far. These communities are at the forefront of the change that is required and they are proactively securing the future for the inhabitants of their village, town or neighbourhood. They are in essence building resilience into their way of being.

Without local networks and a sense of who to turn to when problems arise, communities can suffer. Where individuals have little idea of who their neighbours are, the ability to come together at a time of crisis is less likely. Our worlds have become less dependent on local interaction, as we engage with other business and interaction elsewhere. At no previous point in history have we become so rapidly disconnected from our neighbours because we are too busy to take the time to build the foundations for our family and local citizenry.

Connecting locally takes a deliberate shift of gear: without doing so we simply whizz on by. It takes time and engagement to discover the value of what is often right in front of us. When we stop and pause for a moment, look to our immediate surroundings and see our world as the place where we live, what do we see? For some it is a personal world where interconnectedness already exists, not only human-to-human but also between human, the earth and other beings. For others it is irrelevant whether other non-human beings, beings that have no rights, function well. Loss of species does not enter into their calculation; basic understanding of the importance of other species to our health and well-being is not high on their agenda. The cross-fertilisation of species, interdependency, an understanding that functioning ecosystems are interrelated, without which our lives cannot function, is a core foundation of our communities. Contributing to the local community can pay dividends – look to Shell's example above: they have invested in it, created a training programme and built a skilled community with jobs. Take that same philosophy of giving back to a local community and think how fast the transition would be when that same company and many others foster skill-building and training for a community who need to transition to a post-oil world. Replace the word 'oil' in the box above to 'renewables' and it gives a picture of what needs to be done.

Transitioning communities are born of the desire to have a better alternative to our current set-up which is fuelled by non-renewable energy, compromised food-stuffs and centralised decision-making. However, a swift transition often proves difficult because of the high cost. Policy that supports damaging solutions is at the heart of the problem. Unlock that door, and transitioning communities can flourish faster than any one solution.

All over the world farmers, especially in developing nations, are trying to prevent biotech companies from replacing common crops with genetically modified, proprietary crops whose seeds cannot be shared and whose ecological effects are a curtailment of life. Environmentalists are trying to protect wilderness areas and win fair compensation for the corporate use of public lands. They are striving to prevent multinational water companies from privatising public water works, privatising land for extractive purposes and privatising their forests for carbon markets. These stories and many similar tales exist by the thousands across the world. The facts may differ for each, but the premise remains the same – the community fighting, often a losing battle, against ownership of the commons.

Scientists are now building shared databases of research.

In so doing, researchers are preventing corporations from patenting basic biomedical knowledge. Academics are bypassing commercial journal publishers and creating open and free journals where articles can be accessed via the Internet. Ordinary citizens are rallying to defend the commons of public space by fighting intrusive commercialism in civic spaces, sports, public schools and personal spaces. Local communities are fighting 'big box' retailers like Tesco and Wal-Mart who are threatening independent businesses and local community projects.

The commons is a vision of a world unfettered by intellectual and physical property rights. It is a vision that is growing daily by the actions of a band of resourceful individuals who collectively make an enormous difference to our perception of ownership – or lack thereof. Often the most innovative ideas and beliefs originate and take shape at the periphery; a well known mantra of permaculture practice is that the most effective activity happens where two different ecosystems meet, be it the edge of a pond and the land or where civil engagement meets political engagement. Attempts to sharply confine the two ecosystems results in limited growth. Creating the free space for new life to grow, however, gives birth to enormous diversity of solutions: in a field the most diverse life is at the border. Where non-intervention has prevailed and nature has been left intact, biodiversity has flourished and balance between the field and the countryside prevails. Apply the same principle to climate negotiations and the same can take place: where civil society meets political leaders, innovation that is happening at the edges can be brought into the mainstream very fast so that the fertile seeds of ideas can take root and flourish.

THE TRANSITION ENABLING ACT (TEA)

The purpose: To enable transitioning communities to have primary decision-making powers premised on their rights as set out below.

The enabling provisions:

1. the right not to be polluted;
2. the right to restorative justice;
3. the freedom of a healthy environment.

The breaches:

4. of the duty of care for community health and well-being; of the duty to ensure ecological justice and provision for future generations;

5. of the duty to place the long-term needs of the wider community above short-term return.

....

CANAL ENABLING ACTS OF THE 18TH CENTURY

Trevithic's first steam train was built in 1804 and Stephenson's first railway line was opened in 1825. Neither would have been able to change the course of history were it not for a simple law which brought about the canals. Canals criss-crossed the country, facilitating the transport of coal and other heavy goods to towns and ports and were a crucial mainstay of the Industrial Revolution. The canals were the highways that brought fuel to the people. Before that, society had little access to coal.

Prior to the canals, the movement of goods to market came at great expense, was dangerous and cumbersome. The primary mode of transport was the horse, but a horse can only carry so much weight on it's back and on a wagon. A canal barge however could carry up to 400 times the weight, with just one horse to pull it along a tow-path.

The opening of canals brought belching fumes; they were to be seen rising above the skyline over towns and cities now that people could afford to heat their homes with coal. A quantum leap in the way goods were produced and exchanged was brought about by the canal system, but what was key to unlocking the door was that it required a simple piece of legislation to prefer the rights of society over the interests of the landowners.

Just one piece of legislation opened the door to a new industrial age which had a huge impact across the world. Speeding the movement of goods (the canals were essentially about moving goods rather than people) had ramifications for all sectors of business that had previously been impossible. Coal could now be transported out of the coal-fields fast. All because of the canals.

It took parliament little time to realise that they had a law – the Canal Enabling Act – that would unplug a seemingly intractable problem: how to override existing property laws to allow the economy to flourish. Ownership of land was preventing the building of canals from the coalfields to the cities. Private ownership was hindering progress. To build a canal needed an Act of Parliament, a law which established that the needs of society came before the interests of landowners. Parliament had the authority to override the interests of the landowners to enable the canals to be built so that the wider interests of society for a richer life became possible.

Thus, the Canal Enabling Acts established a precedent in placing the public interest before the rights of property. Ownership was subject to the requirements of a larger community, namely society at large. The interest of the many was recognised as overriding the interests of the few who owned the land needed for the building of canals. Many thousands of people who had previously lived without, now suddenly had access to cheap coal. Their well-being took precedence over the rights of the few who controlled the land which could now be requisitioned for building the new canals. By the end of the 18th Century over 400 Canal Enabling Acts had been passed for canals that criss-crossed much of the English countryside. Within a short space it became possible to move goods across the country far more cheaply than by packhorse. With the ready availability of coal, the development of the railway system became possible, where again the public interest had to take precedence over the interests of the few.

TRANSITION ENABLING ACTS FOR THE 21ST CENTURY

Just as the British Canal Enabling Acts of the late 18th Century opened the door to society changing beyond all recognition, so too can Transition Enabling Acts be created to enable communities to transition. Enabling Acts are laws that are turnkeys for society: our missing key is a law which can turn our communities around and make them resilient in the face of energy price hikes and shortages.

An Enabling Act's primary purpose is to enable something to happen by cutting through existing laws. Just as the Canal Enabling Acts cut through ownership of land for the benefit of the people, so too can Transition Enabling Acts be created to cut through ownership issues for the benefit of health and well-being of people and planet for our communities who wish to transition fast. By empowering local communities with the overriding proviso that the health and well-being of the wider community comes before all other considerations, law can be an enormous lever for change. By shifting priorities to a system which ensures that all decisions treat health and well-being as number one, cost-benefit analysis comes second to the longer-term vision. Transition Enabling Acts are laws which place health and well-being first. In so doing, countries can build resilient economies.

Transition Enabling Acts (TEAs) can have an enormous impact on procurement decisions at all levels, premised on building internal resilience. It is an Act that looks to the long-term stability of our economies when our global economies begin to falter and it will ensure that we use nature's capital responsibly and with care. This will no longer be 'business as usual' but will be 'business for the future'. To do this will take a brave government which is prepared to look after their people and the welfare of their patch of the planet. No two TEAs are likely to be the same; each nation will have its separate and identifiable set of rules which they believe are most important to their well-being, to best serve their needs and the needs of their land. Both people and planet will be protected by this law. Neither people nor planet will be best served if one is to dominate the other. It is a symbiotic relationship that is sought, where both parties benefit and gain from the other. Without this balance, harmony between the two is not possible and neither will flourish. Planet Earth is a bank whose overdraft facility is about to run dry. Transition Enabling Acts create a new currency, where putting back into the system will be a tool to facilitate investment in the future and secure the best rate of return.

The Enclosure of the Commons

The enclosure of the commons is in essence the privatisation of that which is not ours to own. Markets created to plunder our common wealth has become one of the biggest consequences of globalisation and the neoliberal agenda. Now we have markets which threaten to destabilise our shared interest in the health and well-being of the planet and all who live here.

Companies are taking valuable resources from the commons – spectrum, raw materials, deep-sea minerals, genetic code, public lands, and much more – and privatising them. Once the cash value has been harvested from the commons, corporations simply move on to the next site. What is left in their wake is much damage and destruction. Without rules to determine what happens next, the waste becomes a social disruption. Known to economists as a 'market externality', the waste is abandoned for others to restore.

Enclosure of a commons changes the mindset of a community from a shared benefit to one of private ownership and control. This, in turn, changes our attitude towards and management of the resource, because it becomes a marketable commodity which can be bought or sold just like any other.. This necessitates a different set of rules and different social structures, which realign our dealings with each other and a given resource. Enclosure turns us into a mass of competing consumers, ignoring our shared dependence on the natural world.

As a result of this shift from stewardship and trust to ownership and control it is harder for us to behave as people with commitments to the wider Earth community's health and well-being. Issues which are marginalised by society's norms, such as climate change, take a very distant second place to immediate and individual consumer needs.

In the wake of the financial crisis, calls for transformation can however now be heard. The United Nations Framework on the Convention of Climate Change (UNFCCC) has been called upon to implement a workable, realistic pattern of climate investment that guarantees public control of the global financial system. Other proposals have also been mooted: the World Bank has been called upon to actually implement its review panel's recommendations to stop investing in fossil fuels. Many other proposals are gaining currency. They include calls to remove shareholder primacy; reverse limited corporate liability; refuse intellectual property rights to ideas and innovations that would be best held in common; enable workers and farmers to participate in management decision-making; impose maximum salary limits; withdraw banks' right to value their 'toxic' assets and the call to provide public purse handouts when the public agree.

One thing is certain; we are living in an era of transition on a scale that is unprecedented, the full extent of which we shall not comprehend until we look back in time and say 'this was the beginning of the end of the era of ecocide'.

Chapter 8

OWING A DUTY
OF CARE

'I have a dream.'

THESE were the immortal words of Martin Luther King, spoken on 23 August 1963. It was a statement that embodied a new worldview, one based on equality for all regardless of skin colour. It encapsulated the hopes and aspirations of millions of Black Americans and subsequently many millions more around the world. Those words did more to shape the future for ethnic minorities than any other. Martin Luther King was adamant that equality should never be argued on an economic imperative: he understood to do so would simply commoditise his fellow beings. Similarly, William Wilberforce argued for the abolition of slavery on the basis of the ethical wrongness of treating blacks as property. The end of apartheid was also argued from a belief in its inherent wrongness. Each time the moral imperative trumped the economic imperative.

View the planet as an inert thing and it can be commoditised without a second thought. After all it is just a thing which can be traded and given a monetary value. The Earth becomes a piece of property to trade, a commodity for which a price can be negotiated. Man's dominion over land – his right to extract, pollute and diminish the natural resources of the planet as he wishes – accelerates the asphyxiation of life.

Environmental control has informed our international environmental legislation almost exclusively since the 1970s. Permit allocation, permitting pollution on a limited basis, simply allows the pollution to continue – and often at an increased rate. Permits are allocated, business grows, more permits are handed out, and so what is deemed acceptable expands despite the fact that the pollution is escalating. Businesses that damage the planet continue apace. By commoditising the planet, issuing polluter permits and normalising the process, damaging business practices are legally protected.

Current national and international environmental regulatory frameworks silently reinforce customary rights for business – the right to emit, the right to be inequitable, the right to destroy. The Kyoto Protocol is a document that actively facilitates trading on these terms (permits to pollute, carbon trading mechanisms). Thus an international business has been created to address one symptom (the escalation of greenhouse gases) rather than the problem (the ecocide of the planet – death by a thousand cuts each day). Eighteen years on from the very first climate negotiations and we know that both the mechanisms applied, in microcosm and macrocosm, and the ascribed rights have comprehensively failed to stop damaging practices. Instead, we have enslaved the planet.

The next steps are to re-align our legal structures to stop the damage. Without this happening we have nothing else to secure our future. It's as simple as that. There is no next time. This is the moment we stand up and take action.

Change will come regardless: our lives are continually evolving, as is our universe. Leadership that is in service to the Earth is the new paradigm; leadership which is driven by wider concerns is leadership which puts the Earth community first. Unlike command and control leadership which places power with just a few at the top of the chain of command, service leadership is a relationship-led, a holistic and a non-controlling role. It is an inter-dependent relationship which values both the people in the company and those outside (both human and non-human) who are directly and indirectly affected by decision-making. It is a recognition that there are many stakeholders, not just shareholders, and that the role of a leader is to be the bridge between all of those groups. Decisions which have an adverse impact on any of the wider stakeholder groups often indicate that a wrong turning has been taken. The primary stakeholder is the Earth.

....

MOCK ECOCIDE TRIAL

Supreme Court
London

R v Bannerman & Tench
Law of Ecocide proved at London's Supreme Court:
Two CEOs found guilty on two counts.

A mock trial was held in the UK's Supreme Court on 30 September 2011. The court was packed; press, public and lawyers jostled to listen to the case. Two CEOs had been charged with three counts of ecocide; two counts for ecocide of the Athabasca Tar Sands and one for the ecocide of the Gulf Oil Spill. All evidence was based on real events. The day was streamed live by Sky News to thousands across the world who tuned in to watch the drama unfold. Leading barristers, Michael Mansfield QC and Chris Parker QC, and their teams of lawyers fought for and against the indictments laid at the feet of Mr Bannerman of Global Petroleum Company and Mr Tench of the Glamis Group. Although the CEOs were actors and the companies fictional, the issues the same as if the actors had been the real CEOs of companies involved in the events that were being examined.

Counts 1, 2 and 3 are all set out in Appendix 1. As the trial was listed for only one day, issues which were most likely to arise in a real trial of ecocide, were given priority. The jury acquitted on one count, and convicted the CEOs of Counts 2 and 3. The foreman of the jury in the press conference after the event commented: 'We reached unanimous verdicts on the two tar sands cases very quickly. It seemed to us beyond doubt that the two CEOs had wilfully caused to be created large areas of water that were extremely hostile to life, and would probably remain so long after their companies had left the area. With regard to the oil spill in the Gulf of Mexico, we all agreed that the CEO of Global Petroleum was very guilty of something very serious – but was it ecocide? We had only half an hour to consider this question and, on the evidence presented to us, nine of us were sufficiently unsure to acquit him.'

The results were telling: the issues of size, duration and impact were what the cases examined. The failure of the duty of care owed by those in a position of superior responsibility was conclusive.

Mapping the world's ecological footprint country by country gives a flawed picture. Ecological creditors and debtors are not countries, but communities and a business is a community of sorts. Of course, companies are constantly in flux, changing in size, composition and location. Some communities are ecological creditors; others are racking up an enormous ecological debt. Fracking, mining, unconventional tar extraction, intensive monoculture, genetic modification and terminator seeds, agrochemical inhibitors – these and more are the modern world activities of some of our ecological debtors. To be serious about stopping our escalating greenhouse gases, which are a symptom of our plundering, we now know that we will need to stop the debt from accruing.

LEADERSHIP VACUUM

Commerce has created a form of leadership which places the profit motive above what is for the best in the wider context. It has caused us to lose sight of the moral dimension. Governments who prioritise GDP are caught in a trap of enormous proportions: the assumption that rising numbers equate with increased well-being for all and that faster growth represents greater prosperity, ignoring the risk that overheating may lead to a crash. Ironically, the higher our GDP, the more it is celebrated as an indicator of our nation's wealth. But it is a flawed system: it fails to look at true wealth. A nation where money is regarded as the primary driver will find it difficult to adjust to a world where other values come into play. Gross National Happiness (GNH) measures quality of life or social progress in more holistic and psychological terms than GDP. Although it is a measure decried by those who place their belief in GDP, more and more people are self-examining their own GNH to see what it means to them to find happiness, rather than wealth, in their own lives. Measuring happiness begs more than the question, 'What makes me happy?' It does much more than that; it subtly shifts values from monetary based to inherent wealth – wealth of the joy of life and all of its experiences.

Governments that use GNH rather than GDP as their indicator of how their society is measuring up have a far more holistic approach to business and decision-making. They are no longer driven solely by economic considerations. What is given priority is the well-being and health of the people and their surroundings. Well-being provisions for all trump profit for the few when the health of the community is given priority. Where the long-term is sacrificed for profit, the health of both people and planet suffers.

All of this takes leadership of another kind; leadership that is in service. A vacuum has been created; a space has opened up for new leadership that can adapt fast and look to the long-term well-being of others, not leadership that is beholden to the purse strings of others. Adaptive leadership is premised on the willingness to make decisions from a starting-point that places people and planet first. Decisions made from a premise of being in service brings with it an acceptance that each of us has the right to be heard and that life itself is the goal.

It is so easy to fall back into decision-making that is command and control led. By putting safeguards in place, laws that realign our priorities, we can make it the easy option to remain in service to our Earth community. Where good service leadership is in place, it carries through to home-life. Compare a man or woman who goes home at the end of a working day feeling valued for their work rather than being treated as a mere cog in a wheel. The former is far more likely to bring their well-being into the home than the latter, who will be struggling to find it elsewhere. Their family will reap the benefits as well. The wider social implications of service leadership are easy to see: someone who feels valued in the workplace carries with them a greater sense of well-being which then manifests itself in greater productivity and a sense of belonging. A company is a community, and the greater the cohesion between all members of the community, the greater the success is for the whole.

Superior Rights: Native Title

Ngati Apa[1] was a landmark New Zealand case which demonstrated the right of indigenous cultures to assert native title as a superior right. The doctrine of superior native title states that indigenous persons of a given territory are entitled to have the final say as to what happens to their territory. In the event that the community refuses to release their territory to a company or the government, it remains with them.

Just as there are superior responsibilities, so too there are superior rights, rights that override other existing rights. Native title is one example. It is the overarching right of indigenous peoples to their homeland, which persists despite later colonisation or claims to their territories. Most countries have failed to uphold native title and few know that it even exists as a principle of law. Rarely does a case reach the courts on this issue. Thus, many indigenous communities are unaware that their superior right is recognised in law.

Ngati Apa provides an important precedent for indigenous peoples to argue the validity of their right to native title. That right supersedes subsequent corporate rights over their territories. As the case of Ngati Apa demonstrates, the imposition of another body of law does not in itself extinguish pre-existing customs. Ngati Apa is not alone in this ruling: a growing body of judicial rulings elsewhere echo the same premise: indigenous rights with regard to land title take precedence over later claims made by others to the territory they inhabit.

---

1   Attorney-General v Ngati Apa [2003] 3 NZLR 643

Discrimination is a term we normally use in the context of the workplace. However, it is not confined to such contexts. It can and does apply to many other situations. Ignoring native title amounts to a discrimination against indigenous people. Although discrimination laws are predominantly used for law in the world of business, when we view the Earth as our business, then arguably discrimination laws could also apply. When a government or a company overrules indigenous native title, this can also be challenged as a form of discrimination.

Discrimination laws identify four main areas of employment discrimination; age, sex, race and disability. One type of discrimination is 'indirect discrimination': companies have an explicit duty to make 'reasonable adjustments' where an act can or does discriminate against a person or group of persons. For instance, a company discriminates against a disabled person on the grounds of their disability where the company requires the disabled person to work in a building without a lift, and where he or she would be suitable if there was a lift. Where the company refuses to install a lift, or make any other suitable alternative arrangements, the disabled employee is put at a disadvantage and the company has indirectly discriminated against the employee by failing to make reasonable adjustments. The company is obliged to make sure that all reasonable steps are taken so that the employee is not disadvantaged by the requirement that he/she must work in a building without a lift.

The basis of this is that the employer and the employee have entered into a contract, where for receipt of money the employee will undertake certain work. Where the employee is unable to enter into the contract due to lack of reasonable adjustments being made, the contract is breached and the employee can claim damages at an employment tribunal for discrimination on the grounds of failure to make reasonable adjustments.

Likewise, where a company has a contract to use land that is subject to native title, the company can be guilty of discrimination against the indigenous persons on the ground of their native title when the indigenous persons would or could comply with the terms if the company made reasonable adjustments for them.

Reasonable adjustments in this context could include re-housing and providing protection of habitat, as well as compensation. In the event that the community are happy to cede the land on the premise that reasonable adjustments are arranged so that they are not disadvantaged by the loss of land, no discrimination can be said to occur.

A company discriminates against indigenous persons on the ground of their native title when they require the indigenous community to give up some of their land without reasonable adjustments being made. Just as it can be argued that the reasonable adjustment obligation applies to those in a workplace, so too is the obligation owed to those stakeholders on whose land the corporate activity takes place. The company is in effect obliged to make sure that all reasonable steps are taken so that the indigenous community is not disadvantaged by the requirement that they cede their territory to the company. However, where the company has refused to re-house the community, re-forest the land, remediate the waterways, pay compensation etc., the indigenous community is put at a disadvantage and the company has indirectly discriminated by failing to make reasonable adjustments.

Omission by the company to make reasonable adjustments, which result in those with native title being disadvantaged, is a breach of their duty of care and discriminates against those who have the over-riding right to native title. The over-riding obligation is to make reasonable adjustments which remedy the disadvantage. Land that once belonged to others gets passed through many hands; some lawyers argue that it has been divided up so many times that tracing the rightful owners would be pointless. This misses the point – those with native title were there long before the law of contract allowed governmental and corporate acquisition. Just because the land of indigenous people is not codified in a document does not extinguish their right to it. Put another way, a corporate (and often also a governmental) duty of care is owed to indigenous persons who have native title to a territory and who have been or are at risk of being disadvantaged by corporate activities.

SELF-DETERMINATION

The 'open' community of today, replacing the 'closed' community of earlier times, has as its essential characteristic the right of self-determination by its peoples. This right to determine their own fate has become the key, the instrument, for creating an open society. Thus self-determination is a precondition for the very existence of this type of international community. It determines the nature and essence of the present-day international order. It is the, primary principle underpinning the present-day international community – without self-determination, there would be no present-day open international community. Thus, in the hierarchy of the norms of international law, self-determination is an essential first and primary condition from which flow the other principles governing the international community. Self-determination thus belongs to *jus cogens*.

The 'right to development' flows from this right to self-determination and has the same nature. There is little sense in recognizing self-determination as a superior and inviolable principle if one does not recognize at the same time a 'right to development' for the peoples that have achieved self-determination. This right to development can only be an 'inherent', 'built-in' right forming an inseparable part of the right to self-determination.

It will be seen that this might provide a more fertile and more fully explanatory basis for the right of each people and of each State to development. It will be noted, moreover, that the direct descent that we have demonstrated of the right to development from the rootstock of the right to self-determination makes the former much more a right of the State or of the people, than a right of the individual, and it seems to me that it is better that way.*

---

*   Mohammed Bedjaoui, former President of the International Court of Justice, in *Crumbling Foundations; How Faulty Institutions Create World Poverty*, by David Smiley with H. William Batt and Clifford Cobb, 2010 page 215.

NATIVE TITLE IN AUSTRALIA AND THE INTERVENTION

Native title is also known as aboriginal title in Australia. Established Common Law holds that the land rights of indigenous peoples to customary tenure are superior to claims of sovereignty. Each jurisdiction has its own way of determining recognition of aboriginal title, the methods of extinguishing aboriginal title, and the availability of compensation in the case of extinguishment. Nearly all Common Law jurisdictions are in agreement that aboriginal title is inalienable, except to The Crown, and it can be held either individually or collectively.

Aboriginal title was first accepted in the early 19th century. However, it is only within the past twenty years that aboriginal title litigation has resulted in victories for indigenous peoples. Many cases have been brought to trial in Australia, Canada, Malaysia, New Zealand, and the United States and a number are being cited as persuasive authority across jurisdictions. Many now believe that the indigenous right to native title is applicable in all Common Law legal jurisdictions. In 2008 Kevin Rudd, then Prime Minister of Australia, gave his famous message of apology to the Aborigines, which should have been a signal for the shift in recognition of native title.

However, just a few months earlier the Australian Federal Government had taken a step backwards when, in 2007, they had announced their plan to introduce the Northern Territory Emergency Response (NTER), or 'the Intervention' as it is more commonly called. The Intervention had come just six days after the release of a seminal report called *Little Children are Sacred*[2] which called for community governance in Aboriginal communities. Instead, the army was sent in to control land and people. The reason given was child abuse. To control the people, 'special measures' were introduced and the Racial Discrimination Act was suspended pending satisfactory control of various Aboriginal communities. In June 2011, a second consultation was proposed by the government after four years of Intervention which has

---

2 Pat Anderson and Rex Wild QC, *Ampe Akelyernemane Meke Mekarle: Little Children are Sacred* report, 15 June 2007. The report brought to attention to alleged serious problems of sexual abuse, and other abuse, of Aboriginal children and highlighted the failure of Australian governments over decades to provide basic services to address the growing problems in the areas of health, rehabilitation, family support services, empowerment of aboriginal communities and the appointment of a commissioner for children and young people.

primarily served to damage and destroy people's spirit, despite no formal evidence being presented in court to justify the decision. The Australian government have taken the added precaution of preventing anyone from taking litigation by suspending their race discrimination laws. The Intervention is Australia's biggest and most hidden ecocide.

Cultural genocide, a term often used by indigenous communities that have had their rights withdrawn, is a form of ecocide. It may not be as apparent as the overt damage and destruction of land. However, a community is an ecosystem in its own right and where the community has been subjected to intervention which leads to a severe diminution of their ability to function, an ecocide can be said to have occurred. For the many Aboriginal communities in the Northern Territory the ongoing Intervention has caused ecocide that has destroyed their culture and their native rights to their land. In an echo of past colonial times, the Aborigines have suffered the indignity of having their welfare monies replaced with a 'Basics Card' income management, compulsory expropriation of land by the government and other restrictive and discriminatory law enforcement measures.

On 21 June 2010 the Australian Parliament passed the Social Security and Other Legislation Amendment (Welfare Reform and Reinstatement of Racial Discrimination Act) Bill 2009. Instead of reinstating the Racial Discrimination Act, in a step that was to prove even more restrictive, Parliament extended the remit of income management to all welfare recipients in the Territory. By extending the provision to all it could, they claimed, no longer be called discriminatory. However, given that most recipients in the territory are Aboriginal, discrimination is continuing despite recommendations that it be removed. The continuation of compulsory short-term leases and other 'special measures' remain firmly in place. In short, the Aborigines of Australia have lost in one fell swoop the control over their land and their communities: their right to native title and right to self-determination has been extinguished. Australia has faced questioning at the United Nations by member states and independent experts. Failure to repeal the Intervention has lead many lawyers and judges to speak out against it and a 'Will of the People' letter was submitted by Rev Dr Djiniyini Gondarra OAM on the 26 June 2011. The letter, which was a response to the Prime Minister Julia Gillard's announcement of a Second Intervention in the Northern Territory, set out clearly the will of the people to have the Intervention removed, that the name 'Intervention' and 'Emergency Response' be removed from any further initiative and that it be replaced with appropriate support and beneficial development in the Aboriginal communities to maintain their indigenous languages, cultural practices and the capacity to live and work on their land.

Under the Australian Constitution, the provision for a 'Will of the People' letter exists for the people of a given community. It requires a 51% majority vote to establish the proposal as the will of the people. Such a letter must by law be honoured by the government. It is a legal mechanism that may just prove to be the key to unlocking the door to healing this particular ecocide. The government's response has yet to be heard.

Cases in other countries are often instructive; in 1921 a landmark customary land title case was heard. *Amodu Tijiani v Southern Nigeria*[3] upheld a customary land claim which set the precedent that each case must include the 'study of the history of the particular community and its usages in each case'. Subsequently, the Privy Council has issued many opinions confirming the existence of aboriginal title. The test for native title can be found in the more recent case of *Delgamuukw v. British Columbia* (1997)[4]:

in order to make out a claim for aboriginal title, the aboriginal group asserting title must satisfy the following criteria:

---

3   Amodu Tijani v. Southern Nigeria (Secretary) [1921] 2 AC
4   Delgamuukw v. British Columbia [1997] 153 D.L.R. (4th), and see Mark Walters, 'Mohegan Indians v. Connecticut (1705-1773) and the Legal Status of Aboriginal Customary Laws and Government in British North America', *Osgoode Hall Law Journal* 33:4 (1 January 2007)

i. *the land must have been occupied prior to sovereignty,*
ii. *if present occupation is relied on as proof of occupation pre-sovereignty, there must be a continuity between present and pre-sovereignty occupation, and*
iii. *at sovereignty, that occupation must have been exclusive*

Where it is accepted that a people of a given territory have native title to land, the inevitable step is to accord native title to the land itself and view it as having intrinsic rights as well. Many indigenous communities adhere to unwritten laws of nature which are premised on the existence of rights for trees, birds, animals, water, fish, soil to exist. Often referred to as 'lore', it is their law and is equally valid, if not more so than recent written documents that the non-indigenous world uses.

Governmental Interventions, such as the one imposed on the Northern Territory, is in direct contravention of existing native title case law. It is in contravention of many other non-binding treaties and declarations, including their own natural laws. Yet, governments have ridden rough-shod over the lives of many people. Morally it is a crime against humanity, a crime against nature, a crime against future generations and a crime against peace.

The scales of justice today have become unbalanced. Now there is a growing awareness that the law has become too heavily weighted in favour of protecting the rights and benefits of the few, with property and contract laws used to justify the right of a few to commoditise, use and abuse without enforcement of any legal duty of care. The rebalancing of the scales requires a gentler approach, one in tune with an intrinsic valuing of the wider earth community. This would mean the use of trusteeship laws, to realign our roles as guardians or stewards. Through the use of our language we bring to the collective consciousness that which needs to be given name, that which morally we can no longer ignore. In identifying the behaviour of those who are destroying the Earth as perpetrators of ecocide, we can shift to stewardship as the new norm. Just as we reached a point where human life was finally deemed sacred, so now we are waking up to the same thought for the Earth.

Chapter 9

THE SIGNIFICANCE OF LIFE

IF current trends to persist with 'business as usual' continue, they will have dire consequences for climate change, warn the International Energy Agency (IEA). Where there is failure to prevent the increase of greenhouse gases in the atmosphere by reducing emissions, ambient air quality will deteriorate which in turn will cause serious public health and environmental impacts, particularly in developing countries. The IEA is concerned that following this path, 'puts us on a course for doubling the concentration of those (GHGs) in the atmosphere to around 1000 parts per million (ppm) of carbon dioxide-equivalent by the end of this century. This would entail an eventual global average temperature increase of up to six degrees centigrade.'[1]

Global greenhouse gas emissions could rise 50% by 2050 without more ambitious climate policies, as fossil fuels continue to dominate the energy mix, the Organisation for Economic Cooperation and Development (OECD) announced on 15 March.

'Unless the global energy mix changes, fossil fuels will supply about 85% of energy demand in 2050, implying a 50% increase in greenhouse gas emissions and worsening

---

1   International Energy Agency, World Energy Outlook 2009 Fact Sheet
     http://www.iea.org/weo/docs/weo2009/fact_sheets_WEO_2009.pdf

urban air pollution,' the Paris-based OECD said in its
environmental outlook to 2050.

'The global economy in 2050 will be four times larger than
today and the world will use around 80% more energy. But
the global energy mix is not predicted to be very different
from that of today', the report said. Fossil fuels such as
oil, coal and gas will make up 85% of energy sources.
Renewables, including biofuels, are forecast to make up
10% and nuclear the rest. Because of such dependence on
fossil fuels, carbon dioxide emissions from energy use are
expected to grow by 70%, which will help drive up the global
average temperature by three to six degrees centigrade by
2100 – exceeding the warming limit of within two degrees
centigrade agreed to by international bodies.

Global carbon dioxide emissions from energy reached an
all-time high of 30.6 gigatons in 2010 despite the economic
downturn, which reduced industrial production. The
financial cost of taking no further climate action could result
in up to a 14% loss in world per capita consumption by
2050, according to some estimates. Human costs would also
be high as premature deaths from pollution exposure could
double to 3.6 million a year, the OECD said. Demand for
water could rise by 55%, increasing competition for supplies
and resulting in 40% of the global population living in water-
stressed areas, while the number of plant and animal species
could decline by a further 10%, according to the report.

To prevent the worst effects of global warming international
climate action should start in 2013. The report recommends
a global carbon market be set up, the energy sector
transformed to low-carbon and all low-cost advanced
technologies should be explored, including biomass energy
and carbon capture. Failure to do so will make it more
difficult to meet the two degree centigrade limit and will

require very rapid rates of emissions cuts after 2020 to catch up. Scrapping inefficient fossil fuel subsidies are essential for renewables growth — which could increase global real income by 0.3% in 2050, the report said.[2]

The writing is on the wall. We have the knowledge that fossil fuel use is putting humans at risk of catastrophic climate change; the OECD report is evidence that cannot be ignored. The current trajectory has enormous consequences: increase in temperature that puts all of life at risk. The use of fossil fuels is no longer an option. Fossil fuel is a dangerous industrial activity for the whole of civilisation. To continue is a miscarriage of justice.

All States have a legal duty of care to prevent loss of life. Where humanity at large is placed at immediate and real risk, emergency measures are called for.

INTERNATIONAL TRUSTEESHIP SYSTEM

'The United Nations shall establish under its authority an international trusteeship system for the administration and supervision of such territories as may be placed thereunder by subsequent individual agreements. These territories are hereinafter referred to as trust territories.'*

The UN Trusteeship Council was a founding pillar of the UN in 1945; its purpose was to assist non-self governing territories after the second world war when colonies were being disbanded. Under the Charter, the Trusteeship Council is authorized to examine and discuss reports from the Administering Authority on the political, economic, social and educational advancement of the peoples of Trust Territories and, in consultation with the Administering Authority, to examine petitions from and undertake periodic and other special missions to Trust Territories. Former colonies were named non-self governing territories (NSGT's) because it was recognized that these were territories that were unable to self-govern and required a period of transition until they were able to self-govern again.

---

2   http://www.oecdbookshop.org/oecd/display.asp?sf1=identifiers&st1=972012011P1

The UN Trusteeship Council was closed in 1994. However, it still exists (it cannot be removed as it is a founding pillar of the UN). The role of the Trusteeship Council was to give assistance to territories that have become non-self governing (after former colonial power had withdrawn). Ecocide can render a territory non-self governing overnight. With the threat of rising sea-levels, melting ice and floods, the Trusteeship Council provides an existing mechanism to house Member State representatives to determine how to discharge their legal duty of care to assist territories that are at risk or have become non-self governing and ensure that the best interests of the inhabitants are placed first.

---

* Article 75, Charter of the United Nations, 1945

....

Care implies feeling, a word that is rarely used in the marketplace of traders of the earth. 'Ecosystem services' are creating an even larger disconnect from the sense of duty and responsibility that is needed to shape our approach to the way we move forward.

Implementing the moral imperative is the responsibility of all our leaders, and establishing a new ethical business culture is key. Complementary to this, to ensure ethically driven decisions are made by CEOs and directors, are the regulations and tools to implement the values and mission of a company. Co-operation and trust form a bedrock from which transparency of business practice can grow.

Decisions are made from a place inside each of us that holds our values and which then determines the outcome for us – and those affected: we can either take life for granted, something to be enjoyed without thought for the consequences to others or the environment, or we can appreciate its intrinsic value as that which gives us and all around us life and recognise our responsibility to encompass the concerns of the wider community.[3] A prerequisite for effective management and decision-making is for leaders to take time to reflect, and to acknowledge that central to all their

---

3 Kenneth Goodpaster, *Conscience and Corporate Culture* (2007)

business dealings must be a code of responsibility towards people and planet as well as to their shareholders.

Corporate social responsibility (CSR) is an extremely well established term in the business world. It has become a vehicle for promoting transparency, accountability, integrity, better communication, mutually beneficial exchange and trust. In providing a language and vocabulary to critique business from both inside and outside its boundaries, it has become a byword for business ethics and modern capitalism. It is however, without basis in law. Corporations are free to say and do as they wish with little governance other than their own self-appointed auditing.

Voluntary mechanisms that impute intrinsic values on corporations take time and rarely change the *status quo* overnight: only law can do that. The danger always with voluntary initiatives, no matter how well-meaning, is that they become adjuncts, tick-box exercises and veils that hide ever more destructive practices that happen out of sight and out of mind. Voluntary means just that – an act of free will. But where free will clashes with existing legal obligations, such as the supremacy of profit, free will is in truth fettered and compromised, no matter how much a person wants to act freely and in the best interests of others.

Our global voluntary mechanisms are hindered by the law of profit. Corporations are legally obliged to treat profit as their number one fiduciary duty to their shareholders. Attempting to square the circle is just not possible until we re-align our laws with our re-alignment of our values. When we value life itself in law, companies can and will emerge like butterflies out of the chrysalis to become the new life creators rather than the destroyers that they are today.

THE LUNGS OF THE EARTH

The Amazon forest, due to the impact of deforestation, commercialisation and climate change, is already undergoing a widespread dieback, with parts of the forest moving into a self-perpetuating cycle of more frequent fires and intense droughts leading to a shift to savanna-like vegetation. While there are large uncertainties associated with these scenarios, it is known that such dieback becomes much more likely to occur if deforestation exceeds 20–30% (it is currently above 17% in the Brazilian Amazon). It will lead to regional rainfall reductions, compromising agricultural production. There will also be global impacts through increased carbon emissions, and massive loss of biodiversity.

The build-up of phosphates and nitrates from agricultural fertilisers and sewage effluent can shift freshwater lakes and other inland water ecosystems into a long-term, algae dominated (eutrophic) state. This can lead to declining fish availability with implications for food security in many developing countries. There will also be loss of recreation opportunities and tourism income, and in some cases health risks for people and livestock from toxic algal blooms. Similar nitrogen-induced eutrophication phenomena in coastal environments lead to more oxygen-starved dead zones, with major economic losses resulting from reduced productivity of fisheries and decreased tourism revenues.

The combined impacts of ocean acidification, warmer sea temperatures and other human induced stresses make tropical coral reef ecosystems vulnerable to collapse. More acidic water – brought about by higher carbon dioxide concentrations in the atmosphere – decreases the availability of the carbonate ions required to build coral skeletons.*

---

*   http://gbo3.cbd.int/the-outlook/gbo3/executive-summary.aspx

. . . .

Loss of the Amazon means quite literally we lose the very lungs of our Earth. By ignoring the mass slaughter of one of the world's most precious assets in search of profit is to be wilfully blind to the fact that human life is utterly dependent upon its survival. We lose the world's lungs, and we will all find our own ability to breathe easily is lost. The decisions of a very few people lie at the heart of whether we all survive this particular journey. The good news is that the rules of the game can be changed easily and the evolution of law points in the direction that the next stage of the game is ready to take place.

The preamble to the World Charter for Nature (1982) indirectly established the non-human right to life through the acknowledgment that 'life depends on the uninterrupted functioning of natural systems'. Similarly, a number of instruments highlight the dependence of health on the environment. Resolution 23.61 of the World Health Assembly, for example, recommends 'the establishment of effective control over the condition of the environment as a source of health and life for present and future generations'.

The emerging rights of a non-human to life has taken time to develop within the context of rights-based law which has focused on the human right to development as a natural progression out of existing human rights. Our unofficial right to a healthy environment (i.e. not yet recognised as a legal right in international law) has gained much support and definition from soft law (unenforceable declarations). The General Assembly Resolution 2398 [XXIII] on technological change and human rights first made the link between environmental impairment and a breach of a human rights. The Stockholm Declaration begins with the words 'Man has the fundamental right to... adequate conditions of life in an environment of a quality that permits a life of dignity and well-being'. What is implicit in these words is the formulation of a human health and well-being provision that is predicated on the health and well-being of non-human life. The Statement of Forests Principles, 'the right to socio-economic development on a sustainable basis', is another early document which alludes to the right of humans to a healthy environment.

But it goes further than this. It is not only our right to well-being and to enjoy peace by living on land that is healthy. It is the right of the land itself not to be polluted. It is not only that we have duties and obligations to fellow humans, but if we are to accord to ourselves rights that encompass the health and well-being of nature to guarantee our best outcome, then we too have obligations to ensure we do not pollute the land in the first place. If we want to live in harmony with our environment, we too have a part to play in the process. We carry duties and responsibilities with us which apply to all beings, not just human beings.

International environmental law establishes that we have a duty of care to our fellow humans to ensure we do not breach a whole list of rights. Where responsibility for the environment is codified by the right to life, a breach of that right can be relied upon to bring a complaint of the loss, damage or destruction of the environment which has caused impairment of peaceful enjoyment. Individual breaches of human rights are usually tried under national legislation which has within it either implicitly or explicitly embedded the rights within their own law. Although the starting point may be the right to life, our laws do not always explicitly state it as a breach of that right. For example, the Common Law crime of murder does not treat it as a breach of our human right to life. However, it is understood to be there all the same.

The Universal Declaration of Human Rights, the International Covenant on Civil and Political Rights, and the International Covenant on Economic, Social and Cultural Rights, along with the United Nations Charter and various regional human rights instruments, all set out the right to life. It must be noted that the Universal Declaration of Human Rights was not intended to be a legally binding instrument when it was originally created. However, very quickly it came to be considered a binding international law through becoming a peremptory norm. No derogation is permitted when a norm becomes accepted as a peremptory norm. The term 'jus cogens' is given to norms that reach that status, and it is generally accepted that prohibition of genocide, slavery, torture, and human rights are all peremptory norms that no person can evade. It has been argued by legal experts that plundering of the earth violates the customary international law of the human right to life on several counts.[4]

The right to life is the basis of all other human rights – it is the most important right of all. Extinguish life and all other rights are extinguished too. The United Nations Charter in 1945 was the

---

4   Paul Sieghart, *The International Law of Human Rights*, pp. 53, 54 (1983); Mark Allen Gray, 'The International Crime of Ecocide', *California Western International Law Journal*, 26, pp. 215 - 71, (1996)

beginning of a global statement of intent for peace which paved the way for the Universal Declaration of Human Rights in 1948, which in turn opened the door to the International Covenant on Civil and Political Rights,[5] the European Convention on Human Rights[6] and the American Declaration on the Rights and Duties of Man.[7] Human rights evolved to include the duty to protect life by implying the right to a healthy environment and the continued coexistence of people and that environment. The right to life now extends beyond national borders and their citizens. In the *Yanomami* case,[8] the Inter-American Commission of Human Rights determined that, because environmental degradation in the Amazon rain forest can violate the right to life, the Brazilian government, in approving development there, violated that right and rights to health, liberty, personal security, residence and freedom of movement. Destroy our planet and we violate our human right to life; destroy the earth and we violate the earth's right to life as well.

THE ECONOMICS OF TAR SANDS

Companies with big tar sands investments are increasingly at odds with a prevailing belief that is evolving within the international business world; namely, a business which remains tied to a profit dependent on low carbon prices, either by capturing it and storing it or by receipt of subsidies, is high risk. But it is not the risk that is at issue here: it is the consequence that counts. Where the consequence is one that places both human and non-human right to life in danger, endangering civilisation across the world, investment can no longer be justified.

Confirmed levels of emissions produced from Canada's tar sands reserves will rest on the rate and method of extraction as

5   International Covenant on Civil and Political Rights, 1966, Article 6(1)
6   European Convention on Human Rights, 1963, Articles 2(1) & 5(1)
7   American Declaration on the Rights and Duties of Man, 1948, Article 1
8   *Yanomami*, Case 7615, *Inter-Am. C.H.R. Res.* No. 12/85, OEA/. Ser.L/V/II.66, doc. 10, rev. 1, 24 (1985)

well as the degree of improvements in technological efficiency and mitigation in the future. In terms of qualifying emissions against the global carbon budget estimated, total emissions from exploiting Canadian tar sands would equate to 21% of total allowable emissions until 2100 – which equals 19 years' worth of the world's carbon budget. On this basis alone, expansion of the most carbon-intensive industries are not an option where governments and private investors are capable of investing in alternatives that are low-carbon. Tar sands expansion represents a threat on various fronts to the local environment as well as the global climate.

Extracting oil from tar sands is a hugely expensive, energy-intensive and destructive process. Canada contains half the world's boreal forest and 11% of global terrestrial carbon sinks. Peatland and wetlands are destroyed to expose the tar sands, deforestation is significant and the loss of the carbon storage value of these areas is extensive.[9] Huge amounts of water from the Athabasca River are being used, and operations are producing enormous tailing lakes. So far these lakes cover an area of 130 square kilometres, with projected growth to 220 square kilometres, filled with toxic waste that is destructive to birds and wildlife. The traditions and health of Canada's First Nations indigenous communities are also threatened by rising levels of toxins that are found in their water and fish. The entering of toxic waste into the food chain is a possible cause of the rising and unusual incidences of cancer reported throughout all the local communities.

Athabascan oil extraction has been given a new spin: it is being promoted as 'ethical oil' to the USA, the oil of choice over and above Middle East conventional oil. In an irony lost on most consumers, the ethics of oil competition are blind to the politically correct sensitivities of how the Middle East may view themselves. As a recession recedes, production costs increase – which then have an adverse impact on profit margins of oil companies. Forecasts for future profitability based on the current cost to the

---

9   Simon Mui, Luke Tonachel, and Elizabeth Shope, *GHG Emission Factors for High Carbon Intensity Crude Oils*, Natural Resources Defence Council, March 2010.

environment are over-optimistic where rising costs and current and future environmental impact have not been factored in.

Low-carbon fuel standards, like the one already in place in California and those being introduced in British Columbia and Ontario, could decimate the market for dirty fuels like tar sands oil. By making it much more expensive to buy fuel that has very high GHG emissions in its production, will place the spotlight on oil from tar sands and identify it as a very expensive alternative to cleaner options. A strong EU Fuel Quality Directive currently under consideration in Brussels, providing the true measure of carbon to car is included, would prohibit producers from selling fuel from tar sands into the EU market. Whether the EU Directive will have teeth remains to be seen.

Volatility of oil prices over the past few years are a structural feature of global oil markets and are likely to continue. Moreover, oil prices and demand look set to peak and trough without enough time to remain stable at high enough levels for tar sands to deliver sustainable profits[10]. Even excluding rising carbon and environmental costs, oil production from the most accessible tar sands is calculated to be profitable only where the oil price stays above US$75 per barrel. Carbon prices are expected to rise and the pressure mounts for more severe caps on emissions. Buying more allowances at higher prices eventually becomes unsustainable and either businesses look to defray their costs elsewhere or go bust. Neither outcome is satisfactory.

THE TRUE ETHICS OF OIL

An inventory of established issues attributed to oil sands developments was aired by aboriginal leaders and elders from across the Mackenzie River Basin in Alberta. They included the abrogation of treaty rights and the declining health of their people, the land and the water, which was now so contaminated that

---

10 *The Viability of Non-Conventional Oil Development*, Innovest Strategic Value Advisors, March 2009.

fishing and drinking of the water was no longer an option. Federal
authorities had been called upon to step up efforts on monitoring
health, reporting back on their assessments, and stemming the
cumulative impacts of the nearby unconventional oil industry
on their traditional way of life. The government of the Northwest
Territories took no action. Decried by the indigenous communities
for three decades, there has been a continued abdication of federal
responsibility and failure to deliver on their commitments under
the Master Agreement. Testimony given by indigenous members
of the community, by people directly affected by the impact of
the oil industry, was first-hand evidence of the loss of their treaty
rights, their health, access to safe drinking water and the local
fishery, including reliance for sustenance and commercial sales.
Of great concern was the impact on their access to traditional
harvesting territories.

George Poitras is a leader of the Mikisew Cree First Nation.
George spoke out against the government's failure to act:

*'The Mikisew Cree have submitted on many occasions
to the governments of Alberta and Canada concerns
regarding the pace and extent of oil sands development.
Unfettered exploitation of oil sands with little or no regard
to the Mikisew Cree's concerns and claims have left the First
Nation to conclude that both levels of government has de
facto extinguished treaty rights of the Mikisew Cree... the
federal government has both the legal tools and the legal
obligation to protect our rights and our health.'[11]*

Recent court determinations have held that a consultation process
requires the Canadian government to directly engage the affected
aboriginal peoples, give due consideration to their concerns and
minimise any adverse impacts on treaty rights.[12] This is not being
done. Nor has any official federal policy on aboriginal consultation

---

11 Testimony, George Poitras, May 12, 2009. see the *Report of the Standing Committee on Natural
Resources, The Oil Sands: Toward Sustainable Development*, March 2007, 39th Parliament, 1st
session.
12 *Mikisew Cree First Nation v. Canada (Minister of Canadian Heritage)*, [2005] 3 S,C.R. 388

ever been implemented. Chief Jim Boucher, Fort McKay First Nation testified:

> 'My members have lost approximately 60% of their trap lines to oil sands development, and 75% of our lands within twenty kilometres of our communities have been mined or approved for mining. Oil sands leases cover almost all our traditional territory and have effectively extinguished the exercise of our treaty right to hunt, fish, trap, and gather... there is presently no cohesive federal or provincial economic, environmental, or regulatory framework or blueprint to address not only the sustainability of oil sands production, but also its cumulative and long term environmental impacts on water, land, air and aboriginal rights.'[13]

First Nations located further downstream and outside Alberta advised that they have never been consulted by the federal government regarding impacts of the oil sands developments; nor had they been included as parties to intergovernmental water agreements for the area. The government of the Northwest Territories had only consulted on their water policy with one First Nation, the Smith's Landing who are located just south of the Northwest Territories border.

Implementation of the federal Species at Risk Act had put a requirement on agencies to take timely action to protect threatened species, in particular woodland caribou, and a number of parallel legal actions have now been filed by some of the First Nations and two NGOs seeking court orders. It's a small chink of light in a sea of toxic tailings and breaches of human and nature's rights.

The right to life is at the very core of all our rights. Without life, nothing else matters. It is literally the basis from which all other rights spring. Our laws are predicated on fostering life, not death – and where we have laws that indirectly create rights that

---

13 Testimony by First Nations leaders and elders in Edmonton, Testimony, Chief Jim Boucher, May 12, 2009.

allow us to destroy, we now have the knowledge that those are laws which cause enormous inequity to grow and spread like a cancer.

Back in 1945 when the United Nations Charter was written, our understanding of our interconnectedness of life was limited: the world was viewed through a human-centric prism. Now we know better; we know that the life of our planet dictates how we can survive as humans and thus the life of others takes on a new significance. 'Ecosystem services' will soon become a term we shelve as we shift to a deeper understanding of the intrinsic value of the living world as a whole, not as a commodity upon which we place a price. That which we grow – the fruits of our labours – can be traded. However we destroy Nature's gift – the soils and the lands and seas out of which all else grows – at the cost of losing our ability to function at all. Commodify ecosystems and all we do is revert back into the old paradigm that created slavery and markets based on disconnect and ownership; entrust ecosystems into the hands of the keepers of life and we can be sure that they will feed us all for many generations to come. The old system is one based on fear of lack, whereas the latter is based on trust in Nature's bounty. One imposes an extrinsic value, the other an intrinsic value. One looks to the short-term profit, the other to the benefit of future generations. The old paradigm is ripe for conversion to a better world.

Chapter 10

THE END OF ASSET STRIPPING EARTH

'Many commentators have called the Cancún accord a "step in the right direction." We disagree: it is a giant step backward. Bolivia is a nation in the global South. This means we are among the nations most vulnerable to climate change, but with the least responsibility for causing the problem. Studies indicate that our capital city of La Paz could become a desert within thirty years. What we do have is the privilege of being able to stand by our ideals, of not letting partisan agendas obscure our principal aim: defending life and Earth. We are not desperate for money. Last year, after we rejected the Copenhagen accord, the US cut our climate funding. We are not beholden to the World Bank, as so many of us in the south once were. We can act freely and do what is right.'

PABLO SOLON, Ambassador of the Plurinational State of Bolivia to the United Nations, speaking at the conclusion of the UNFCCC COP 16 in Cancun, December 2010.

Leading Scientist Warns that Planet Ecosystems are Close to Tipping Point

'A temperature increase of more than two degrees centigrade is likely to push components of the earth's climate system past critical thresholds, or 'tipping points'. The window of opportunity to avert the most serious impacts of climate change is closing rapidly.'

So said Hans Joachim Schellnhuber, director of the Potsdam Institute for Climate Impact Research and chair of the German government's Advisory Council on Global Change, at a high-level briefing at UN Headquarters in New York. While the Arctic sea-ice and the Greenland Ice Sheet are regarded as the most sensitive of such 'tipping' elements, with warming above two degrees, others like the Amazon rain forest and Indian and African monsoon systems could be drastically altered, he said. Even if global emissions were to peak in 2015, the reductions required thereafter to stay below a critical two degrees centigrade threshold increase would be equivalent to a Kyoto Protocol for all countries every year.[1]

....

AMAZON ECOCIDE*

An Ecuadorian Court ruled in February 2011 against Chevron for its destruction of the Amazon and loss of life. The question is, will Chevron pay up and clean up? With a bill of $8.6bn (£5.3bn) in fines and clean-up costs, plus $900m reparations, to the victims of oil pollution that fouled a swathe of Amazon rainforest along the country's remote north-eastern border, Chevron have their work cut out – if they respect the Ecuadorian court ruling. Their lawyers came out fighting, despite the overwhelming evidence against them. Eighteen years on and tens of millions of dollars invested in contesting the

---

* http://www.independent.co.uk/environment/nature/chevrons-dirty-fight-in-ecuador-2216168.html
Guy Adams,16/02/11

---

1 http://www.un.org/wcm/content/site/climatechange/pages/gateway/the-science/pid/4101

lawsuit has left a sense of frustration at losing, which was evident in the statements Chevron made to the press afterwards. Pledges by their lawyers that appeals would be lodged through every conceivable legal avenue, on at least three continents, smacked of blindness in the face of the truth. Both an Ecuadorian and a US appeals court have now upheld the original decision.

The sum is the largest ever levied in an environmental lawsuit anywhere in the world. It is a landmark judgment, not merely in size of penalty, but in the imposition of accountability upon the corporate world for mass damage and destruction to the environment when those in a position of power fail to carry the obligations owed to the land they have despoiled. Representatives of the indigenous villagers who brought the case called the victory a success in the wider battle to hold multinational corporations to account. It was a David and Goliath victory for them and it sent out a strong signal to many others fighting similar battles that accountability has become the new paradigm.

Mr Fajardo, the lawyer for the residents who brought the case on behalf of their community, declared the 188-page ruling, a 'triumph of justice'. His only disappointment was that the level of damages wasn't higher (When Texaco departed from its activities in Ecuador in 1992, it pledged $40m to clean up some of the damage. It was shortly afterwards, that they filed the first lawsuit in New York). Mr Fajardo added: 'Today's judgment affirms what the plaintiffs have contended for the past eighteen years about Chevron's intentional and unlawful contamination of Ecuador's rainforest. Rather than accept that responsibility, Chevron has launched an appeal against the Ecuadorean courts and the impoverished victims of its unfortunate practices.'

The events date back to 1964, when Texaco first entered a partnership with Ecuador's state oil company, Petroecuador, to extract oil from the country's remote Oriente region. Over a period of almost thirty years, independent experts in court testified that the pollution from the oil wells killed at least 1,400 people, caused an increase in local cancer rates and illness. It was, they confirmed, the result of billions of gallons of waste oil being dumped into open pits which contaminated the water supplies which in turn killed farm animals, fouled fishing grounds and damaged crops – all then were eaten by the local population who began to suffer. Fossil fuel extraction, they countered, is a dangerous industrial activity that puts both human and non-human life at risk of injury and harm. The courts have now conclusively agreed.

In 2001, Texaco was bought by California-based Chevron, which became America's second-biggest oil firm but inherited the still-ongoing lawsuit in the process. For the 30,000 indigenous community members adversely impacted, Chevron carry the duty of care owed for the legacy of debt accrued by the actions of their predecessors.

....

Asset stripping our Earth does not work. That much is clear. What is left behind is a legacy of toxic waste, wastelands and barren soils. This is one business that cannot be broken up into pieces and sold off. The planet was here a long time before humans came along and we have no right to claim ownership. Our roles are as stewards of the earth and all who live here. We can profit from all that grows from our hands, not from what we take without return. When the earth has been divided up, and sold off in small parcels, we shall realise that the whole was greater than the sum of its parts.

Keeping the earth whole is what counts: doing that is a job creation scheme in itself. Restoration of depleted lands is one job that will keep us busy for a while. Where land has been destroyed, it takes about seven years for a discernible change to have taken place, seas can take less. Instead of short-term goals we can put in place cycles that are in tune with nature's own systems. Nature can teach us the value of life – money cannot. Money is at the end of the day just a convenience. It is neither good nor bad, it's what we do with it that can lead to problems – or solutions.

KEEPING THE SHAREHOLDERS HAPPY

Climate negotiations have become a business, keeping many in jobs. It used to be that a business was created with its demise in mind; in the US a charter could only be granted to a company for a given time, to ensure the job was done. The laying of railway lines, building of houses, once done and the company was disbanded. The rationale was that it remain in service for the time it would take to complete the job. We have had climate negotiations for eighteen years now, and we are no closer to determining the remit of the job, never mind starting the hard work. The outcome in Durban in December 2012 was a cogent example of a business that was unable to look further than keeping the shareholders happy – the shareholders being those who make money out of the problem. Surely we can only talk of success when our leaders put in place laws to prevent massive damage and destruction (at least, that which we have control over – human caused ecocide) and assist those who are at risk of being most adversely impacted? Currently no-one is employed to speak on behalf of the earth at the climate negotiations.

The climate negotiations, like most businesses today, foster a non-legacy approach. Just as it is the law for CEOs and directors to put profit first and shareholder return as their primary obligation, so too do the negotiations take the same route. It is not surprising that few CEOs are able or willing to change – quite simply, a CEO has no legal obligation to place environmental consideration as their primary duty. Nor is there one enshrined in the Kyoto Protocol: there is simply no mention of Member states having a legal duty to the earth. For the time being, negotiators – like CEOs – are unaccountable to the earth and all who live here: their number one legal duty is to maximize profit.

What I saw in Durban was a new wave of dialogue surging forward: the discourse centred on legacy, leadership and law. There was a new focus on how we create laws to put people and planet first. Morality and law converge when we embed intrinsic values in our business practices: quite literally we make Earth our business. The rules of the game are already changing: instead of rules that foster injustice, new rules that affirm life begin to take shape.

It will take a new kind of leadership: bold, courageous and moral. What is emerging is a need for adaptive leadership, leadership that is governed by intrinsic values and is able to move fast in rapidly changing times. There is a growing recognition that when our leaders act upon their duty to their people, closing the door to injustice will become the new norm. There is, there on the very crest of the wave, a different vision emerging and that is led by adaptive leadership – one in which each and every one of us takes the lead by putting people and planet first. That wave looks set to break over the parapet very soon. When that happens, justice, equality and peace will become the new norm and if ever there was a time for us to be adapting, it is now.

Heroes

In Durban I met with many who agreed that we need a new moral code based on laws for the earth such as ecocide and Earth rights. I met with faith leaders, politicians, youths, people who are already changing their world and their businesses. The heroes of the day were the youths; Canadian youths stood up in front of their Canadian Environment Minister, Peter Kent, when he was addressing the plenary on why Canada has decided to withdraw from the Kyoto Protocol. Six youths stood up, swivelled, and turned their backs on him and revealed t-shirts emblazoned with 'People and Planet before Polluters'. No words could speak louder than their gesture, for which they received a standing ovation from those who witnessed their protest before being marched out and stripped of their right to re-enter the negotiations. The response to the Canadian Six demonstrates the two conflicting approaches; cheers for speaking the truth, then banishment for doing so. Up next was an American youth delegate who rose in front of the hall of delegates to protest that the US negotiators 'cannot speak on behalf of the United States of America'. What she went on to say stopped many in their tracks; not because it was radical but because she voiced what many people were afraid to say. 'The obstructionist Congress has shackled a just agreement and delayed ambition for far too long', she said, '2020 is too late to wait.'

'2020; it's too late to wait' – it was a phrase that had resonance right across the world, and for me summed up eighteen years of negotiations, which look set to keep many in a job for another six years until a decision may or may not be made. Twenty-three years of an industry and nothing to show. This is one industry whose demise is well over-due. 2012 will be the year of the youth; a year when youths all over the world start to speak out and call for a new form of leadership – leadership that all can participate in, one that is in service to the wider Earth Community.

....

THE DURBAN PACKAGE: 'LAISSER FAIRE, LAISSER PASSER'

The Climate Change Conference ended two days later than expected, adopting a set of decisions that were known only a few hours before their adoption. Some decisions were not even complete at the moment of their consideration. Paragraphs were missing and some delegations didn't even have copies of these drafts. The package of decisions was released by the South African Presidency with the ultimatum of 'Take it or leave it'. Only the European Union was allowed to make last minute amendments at the plenary. Several delegations made harsh criticisms of the documents and expressed their opposition to sections of them. However, no delegation explicitly objected to the subsequent adoption of these decisions. At the end, the whole package was adopted by consensus without the objection of any delegation. The core elements of the Durban Package can be summarised as follows:

1) A Zombie called Kyoto Protocol
   - A soulless undead: The promises of reducing greenhouse gas emission for the second period of commitments of the Kyoto Protocol represent less than half of what is necessary to keep the temperature increase below two degrees centigrade.
   - This Zombie (second period of the Kyoto Protocol) will only finally go into effect next year (COP 18).
   - It is not known if the second period of the Kyoto Protocol will cover five or eight years.
   - United States, Canada, Japan, Russia, Australia and New Zealand will be out of this second period of the Kyoto Protocol.

- This will be known as the lost decade in the fight against climate change.

2) New regime of '*Laisser Faire, Laisser Passer*'

- In 2020 a new legal instrument will come into effect that will replace the Kyoto Protocol and will seriously impact the principles of the United Nations Framework Convention on Climate Change.
- The core elements of this new legal instrument are already evident from the results of the negotiations:
  - a) voluntary promises rather than binding commitments to reduce emissions,
  - b) more flexibilities (carbon markets) for developed countries to meet their emission reduction promises, and
  - c) an even weaker compliance mechanism than the Kyoto Protocol.

- The new legal instrument will cover all the States, effectively removing the difference between developing and developed countries. The principle of 'common but differentiated responsibilities' already established in the Climate Change Convention will disappear.
- The result will be the deepening of the '*Laisser faire, laisser passer*' regime begun in Copenhagen and Cancun which risks increasing temperatures by more than four degrees centigrade.

3) A Green Fund with no funds

- The Green Fund now has an institutional structure in which the World Bank is a key player.
- The $100 billion is only a promise and will NOT be provided by the developed countries.
- The money will come from the carbon markets (which are collapsing), from private investments, from credits (to be paid) and from the developing countries themselves.

4) A lifesaver for the Carbon Markets
- The existing carbon markets will live regardless of the fate of the Kyoto Protocol.
- Also, new carbon market mechanisms will be created to meet the emissions reduction pledges of this decade.
- It is a desperate attempt to avoid the loss of the carbon markets, which are collapsing due to the fall in value of the carbon credits, from €30 per ton to €3 per ton of carbon dioxide.
- Developed countries will reduce less than they promise because they will buy Emission Reduction Certificates from developing countries.

5) REDD: a perverse incentive to deforest in this decade
- If you don't cut down trees you won't be able to issue certificates of reduction of deforestation when the REDD (Reducing Emissions from Deforestation and Forest Degradation) mechanism comes into operation.
- CONSEQUENCES: deforest now if you want to be ready for REDD.
- The safeguards for indigenous peoples will be flexible and discretionary for each country.
- The offer of funding for forests is postponed until the next decade due to the fact that demand for Carbon Credits will not increase until then because of the low emission reduction promises.

Amandla! Jallalla!

In the actions and events of the social movements in Durban, two battle cries emerged: 'Amandla' and 'Jallalla'. The first one is a Xhosa and Zulu word from South Africa which means 'power'. The second word is an expression in aymara which means 'for life'. 'Amandla! Jallalla!' means '¡Power for life!' This is the 'power for life' that transcends borders, that we must build, from our communities, neighbourhoods, workplaces and places of study in order to stop this ongoing genocide and ecocide.

These are the conclusions on the final outcome of the UNFCCC COP 17 by Pablo Solón, former United Nations Ambassador and Chief Climate Change Negotiator from the Plurinational State of Bolivia.*

---

* http://pablosolon.wordpress.com/2011/12/16/the-durban-package-laisser-faire-laisser-passer/

....

Private investment in restoration projects are few and far between: questionable marketability, limited benefits which take time to materialise and high costs are the antithesis to private investment. Because of this, government funding of restoration budgets is critical. Government support and coordination of stakeholders is particularly important for mega-sites of degradation with large-scale complex interactions and far-reaching implications. What can be achieved is dependent on government commitment and institutional support. However, voluntary intergovernmental agreements take time to put in place and are often situational specific.

The concept of 'ecosystem services' is flawed in a number of respects. The premise that services from nature can be monetarised to enable profit as a way of ensuring conservation is fallacious. Imposing an economic value and protecting nature are all too often mutually exclusive goals. One clear example of how economics can run counter to the problem is Africa's Lake Victoria. Introduction of the invasive Nile perch (*Lates niloticus*) to boost the economic value of the lake contributed to the eradication of local biodiversity. For the local people profiting from trade in the fish, the perch were a success, whereas biologists condemned the introduction of the ecosystem service as, 'the most catastrophic extinction episode of recent history'.

The logic of ecosystem-service-based conservation rests on the implicit assumption that the biosphere is benevolent. It provides us with useful services. Such reasoning overlooks the fact that environments don't act for the benefit of any single species and it does not factor in malevolent forces such as hurricanes, floods and rising temperatures. The opposite could also be said: there are 'ecosystem disservices'. Trees take water out of watersheds; forests may be contributing to global temperature increases; wild animals kill people and destroy property; and wetlands can increase the risk of disease. Market-based conservation strategies, as currently proposed, fail to consider how we are to protect nature when the interests of nature conflict with our interests or at those areas where there is no discernible pecuniary gain.[2]

INVESTMENT IN ECOLOGICAL INFRASTRUCTURE

Ecological infrastructure refers to nature's capacity to provide freshwater, climate regulation, soil formation, erosion control and natural risk management, amongst other services. Maintaining nature's capacity to fulfil these functions is often cheaper than having to replace lost functions by investing in alternative heavy infrastructure and technological solutions. The benefits of ecological infrastructure are particularly obvious with regard to provision of water purification and waste water treatment. However, despite some impressive exceptions, these kinds of values are often understood only after natural services have been degraded or lost – when public utilities face the bill for providing substitutes.

Risks of natural hazards are predicted to increase with climate change and have significant impacts in some parts of the world. Coastal realignment, storms, flooding, fires, drought and biological invasions could all significantly disrupt economic activity and society's well-being. Natural hazard control can be provided by forests and wetlands (e.g. flood control) and on the coast by mangroves or coral reefs (e.g. reducing impacts from storms and tsunamis)

Ecological infrastructure investments can be justified on the basis of one valuable service but they become even more attractive when the full bundle of services provided by a healthy ecosystem is taken into account. This strengthens the case for integrated approaches to valuation and assessment: considering possible investments from a single-sector perspective may overlook

2   see Douglas J. McCauley, 'Selling Out on Nature', *Nature* 443, 27-28 (7 September 2006) http://www.nature.com/nature/journal/v443/n7107/full/443027a.html

supplementary key benefits.

The spatial dimension of ecological infrastructure – beyond site boundaries to the web of connected ecosystems – needs consideration for similar reasons. When deciding on management actions and investment in a river system, for example, it is essential for coherent management of the river as a whole to look both upstream to the source and downstream to the wetland or delta created. The decision maker needs to take on board that actions benefiting people downstream have to be implemented upstream. This calls for consistent land use planning and collaboration between countries, communities and people throughout the river basin.*

---

*   *The Economics of Biodiversity and Ecosystem Final Report*, D-1, pp 19 – 26 http://www.teebweb.org

....

Looking both upstream and downstream at the problem opens up our understanding of how to change the rules of the game. Stabilisation policy does just that – policy that creates stability both up and downstream. A law that prohibits extensive damage to, destruction of or loss of ecosystems closes the door to the carbon majors, industries that cause extensive carbon dioxide emissions. Acting as a disruptor, the Law of Ecocide will create a new wave of innovation and new solutions that are non-destructive to ecosystems and do not cause extensive carbon dioxide emissions. Cities and countries will benefit from a surge in jobs, investment and money.

A Law of Ecocide imposes an international and trans-boundary duty of care on any person or persons exercising a position of superior responsibility, without exemption, in either private or public capacity to prevent the risk of and/or actual extensive damage to or destruction of or loss of ecosystem(s). A new economic foundation health and well-being will be the new norm. Businesses that are already working successfully in the green sector will be able to expand exponentially, which in itself creates jobs and brings investment flow. New industry will undergo a seismic change, with the help of banks who will invest in business that can demonstrate that they are compliant with their legal duty of care to the earth. Instead of the 'polluter pays,' the new rule 'polluter doesn't pollute' becomes the enforceable determinant.

LAW AS A CONSCIOUSNESS SHAPING TOOL

Law shapes our societies, our way of thinking, our behaviour. By imposing upon nature the concept of it being property, legal systems have legitimised and encouraged the abuse of Earth by humans. Our laws are built on the premise that humans have superior rights to the planet. It is our laws that have given us the right to take and to pollute so extensively, so much so, it is now considered our norm. The concept of Earth as a living organism has been forsaken. The prevailing belief that has wrought our predicament has treated the earth as a mere resource to be plundered at will. As a consequence the imbalance in our ecosphere is now so great that it is threatening to destabilise all of Earth and mankind. We have to redress the imbalance, to ensure the scales of ecological justice are in equilibrium once again. After all, our commonality is the earth we walk on, the soil that feeds our plants, the trees that provide shelter and warmth, the air that we breathe.

Such is the crisis we face that we now recognise the need to protect and ensure that such protection is effective and global in application. For that we need the use and support of laws. Much as a voluntary code is useful and even desirable, past experience dictates a different ending: leaving corporations to implement their own voluntary codes in reality achieves little of true substance. Only by implementing international laws and mechanisms premised on intrinsic values will we embed the recognition of the inherent rights of nature and create the powerful shift in business and consciousness that is vital to turn our world around.

It is up to us to change our legacy to one of ecological credit rather than leaving a legacy of ecological debt. We can each leave a legacy of ecological debt or a legacy of ecological credit. Which is it to be?

It's one thing for society to be saddled with an existing energy strategy that will result in dangerous climate change: it's another thing when existing technologies are driving us headlong into climate disaster – continued use and increased commercialisation of fossil fuel extraction is one legacy of debt we cannot afford to accrue.

There is a plethora of social, environmental and economic reasons why the commercialisation of unconventional oils by the rich developed world is unworkable: at the end of the day the bottom line is that it is wrong in and of itself. It is untenable to expect the global south to adopt a low carbon path when a) the legacy debt of carbon dioxide already in the atmosphere is historically from the global north's industrial activity and b) we proceed by increasing our debt by exploiting even more climate hostile sources of energy.

Making sense of our world takes time. As a civilisation we are now at the brink of collectively taking responsibility for our mess. We know what needs restoring and we know what needs to stop. We have the solutions all around us, now all we need do is take the first step, and very soon we will look back in wonder at all that we have achieved in our lifetime.

We are alive at the most incredible time in the whole of the history of civilisation. For me it is an honour to be here at this time and be a participant on the stage of the new world. You too are on this stage with me, helping me and others to change the systems that are holding us back from co-creating the new world.

THE SOLUTION

Create a green economy. Remove all subsidies for business that contributes to damaging or destructive activity; prohibit all dangerous industrial activity with a Law of Ecocide and put in place subsidies for business that contributes to a low carbon economy.

Transitional provisions will be a necessity. Many companies will require help to transition out of their current business models.

To do that will call for a phased withdrawal of projects that are causing mass damage and destruction. For oil companies, this will mean that their strategies for expanding into renewables will be prioritised and they will require subsidies to help build their infrastructure sufficiently in a five-year time-span.

A Law of Ecocide will end the asset stripping of our future. It creates a level playing field for all. Fast forward the clock a decade: our lands will no longer be shackled by the burden of ecological debt, instead we will have turned our world around. A Law of Ecocide will halt dangerous industrial activity. A law that prohibits extensive damage to, destruction of or loss of ecosystems closes the door to all the carbon majors. The green economy can flourish; prosperity without austerity will become the norm.

One thing is clear; we have entered into a new world. Governments across the world are agreed that an increase in temperature of more than 2 degrees is a danger to human life[3]. The OECD report[4] is our evidence. Our existing course must change. Existing law is no longer fit for purpose. This is our moment of change when we can build a bridge to take us to a future we want; leaders who can adapt will have the capacity to face the rapid changes and make good in times of emergency. Emergency law is what the Law of Ecocide is, and we are living in a state of emergence – of something new and full of the potential to be life affirming rather than life destroying. My book sets out the law to close the door to mass damage and destruction; by setting out in the appendices a sample indictment, the Ecocide Act for all nations to adopt, and the guidelines and new banking rules that all banks can use, we can pre-empt any further ecocide. We have an incredible opportunity now to create a lasting legacy. Together, we can end the era of ecocide.

---

3   ref. the Copenhagen Accord, United Nations Framework Convention on Climate Change 2009, see p.13
4   http://www.oecd.org/document/11/0,3746,en_2649_37465_49036555_1_1_1_37465,00.html, see p.17, 123-125

APPENDICES

SAMPLE ECOCIDE
INDICTMENT

In the Supreme Court
sitting as Westminster Crown Court
30 September 2011
In the matter between

Regina -v- Bannerman & Tench

Indictment

Count 1

STATEMENT OF OFFENCE
Ecocide contrary to section 1(1) and section 2 of the Ecocide Act 2010

PARTICULARS OF OFFENCE

Between the 22nd day of April and 31st day of August 2010 in his role as Chief Executive Officer, **Mr. Bannerman of Global Petroleum Company (GPC)** had authority over and responsibility for a semi-submersible Mobile Offshore Drilling Unit when an explosion on the platform caused an oil spill in excess of 250 million gallons of crude oil into the Gulf of Mexico sea resulting in extensive destruction, damage to or loss of ecosystem(s) covering an area in excess of 200 square kilometers of ocean, to such an extent that the peaceful enjoyment by the inhabitants of that territory has been severely diminished thereby:-

1. causing injury to 2,086 birds;
2. causing the death of 2,303 birds;
3. putting birds at risk of injury.

contrary to section 1(1) and section 2 of the Ecocide Act 2010.

Count 2

STATEMENT OF OFFENCE
Ecocide contrary to section 1(1) and section 2 of the Ecocide Act 2010

PARTICULARS OF OFFENCE

Between 28th day of March and 6th day of September 2010 in his role as Chief Executive Officer, **Mr. Bannerman of Global Petroleum Company (GPC)**, had authority and responsibility of all GPC operations in the Tar Sands. As a consequence of extraction from the Athabasca Tar Sands in Canada the creation of tailing ponds has led to extensive destruction, damage to or loss of ecosystem(s) to such an extent that the peaceful enjoyment by the inhabitants of that territory, and of other territories, has been severely diminished thereby putting birds at risk of injury and death, contrary to section 1(1) and section 2 of the Ecocide Act 2010.

Count 3

STATEMENT OF OFFENCE
Ecocide contrary to section 1(1) and section 2 of the Ecocide Act 2010

PARTICULARS OF OFFENCE

On 19th April 2011 in his role as Chief Executive Officer, **Mr. Tench of Glamis Group**, had authority and responsibility of all Glamis Group operations in the Tar Sands. As a consequence of extraction from the Athabasca Tar Sands in Canada the creation of tailing ponds has led to extensive destruction, damage to or loss of ecosystem(s) to such an extent that the peaceful enjoyment by the inhabitants of that territory, and of other territories, has been severely diminished thereby putting birds at risk of injury and causing 1,600 birds to die, contrary to sections 1(1) and section 2 of the Ecocide Act 2010.

ECOCIDE ACT

Preamble

Ecocide as the 5th international Crime Against Peace

Ecocide is the extensive damage to, destruction of or loss of ecosystem(s) of a given territory, whether by human agency or by other causes, to such an extent that peaceful enjoyment by the inhabitants of that territory has been or will be severely diminished.

The objective and principles governing the creation of the offence of Ecocide as the 5th international Crime Against Peace:

1. To stop the extensive damage to, destruction of or loss of ecosystems which is preventing peaceful enjoyment of all beings of the earth and to prevent such extensive damage to, destruction of or loss of ecosystems from ever happening again.

2. Ecocide is a crime against peace because the potential consequences arising from the actual and/or future extensive damage to, destruction of or loss of ecosystem(s) can lead to:-

   i. loss of life, injury to life and severe diminution of enjoyment of life to human and non-human beings;

   ii. the heightened risk of conflict arising from impact upon human and non-human life which has occurred as a result of the above;

   iii. adverse impact upon future generations and their ability to survive;

   iv. the diminution of health and well-being of inhabitants of a given territory and those who live further afield;

   v. loss of cultural heritage or life.

3. The aim of establishing the crime of Ecocide is to:-
    i.   prevent war;
    ii. prevent loss and injury to life;
    iii. prevent dangerous industrial activity;
    iv. prevent pollution to all beings;
    v.   prevent loss of traditional cultures, hunting grounds
         and food.

4. The crime of Ecocide creates an international and trans-
    boundary duty of care to prevent the risk of and/or actual
    extensive damage to or destruction of or loss of ecosystem(s).

5. All Heads of State, Ministers, CEOs, Directors and any
    person(s) who exercise rights, implicit or explicit, over a given
    territory have an explicit responsibility under the principle of
    superior responsibility that applies to the whole of this Act.

6. This Act places upon all Heads of State, Ministers, CEOs,
    Directors and/or any person who exercises jurisdiction over
    a given territory a pre-emptive legal obligation to ensure their
    actions do not give rise to the risk of and/or actual extensive
    damage to or destruction of or loss of ecosystem(s).

7. The burden of responsibility to prevent the risk of and/
    or actual extensive damage to or destruction of or loss of
    ecosystem(s) rests jointly with any person or persons,
    government or government department, corporation or
    organisation exercising a position of superior responsibility
    in respect of any activity which poses the risk of and/
    or actual extensive damage to or destruction of or loss of
    ecosystem(s).

8. The primary purpose of imposing an international and trans-
    boundary duty of care is to:-
    i.   hold persons to public account for the risk of and/or
         actual extensive damage to or destruction of or loss of
         ecosystem(s);
    ii. enforce the prevention of risk of or actual extensive
         damage to or destruction of or loss of ecosystem(s);
    iii. evaluate consequence of risk of or actual extensive
         damage to or destruction of or loss of ecosystem(s).

9. The offences created under this Act are strict liability; sentence will be determined by the culpability of the person(s) and organisation found guilty as per the provisions of this Act.

10. This Act shifts the primary focus away from evaluation of risk to evaluation of the consequences whereby risk of Ecocide gives rise to the potential for and/or actual extensive damage to or destruction of or loss of ecosystem(s).

11. This Act creates a legal duty of accountability and restorative justice obligations for a given territory upon persons as well as governments, corporations and or any other agency found to have caused the Ecocide.

PART I

Definition of Ecocide

1. Ecocide

Ecocide is the extensive damage to, destruction of or loss of ecosystem(s) of a given territory, whether by human agency or by other causes, to such an extent that:-

(1) peaceful enjoyment by the inhabitants has been severely diminished; and or

(2) peaceful enjoyment by the inhabitants of another territory has been severely diminished.

2. Risk of Ecocide

Ecocide is where there is a potential consequence to any activity whereby extensive damage to, destruction of or loss of ecosystem(s) of a given territory, whether by human agency or by other causes, may occur to such an extent that:-

(1) peaceful enjoyment by the inhabitants of that territory or any other territory will be severely diminished; and or

(2) peaceful enjoyment by the inhabitants of that territory or any other territory may be severely diminished; and or

(3) injury to life will be caused; and or

(4) injury to life may be caused.

Breaches of Rights

3. Crime against Humanity

A person, company, organisation, partnership, or any other legal entity who causes Ecocide under section 1 of this Act and has breached a human right to life is guilty of a crime against humanity.

4. Crime against Nature

A person, company, organisation, partnership, or any other legal entity who causes Ecocide under section 1 of this Act and has breached a non-human right to life is guilty of a crime against nature.

5. Crime against Future Generations

A person, company, organisation, partnership, or any other legal entity who causes a risk or probability of Ecocide under sections 1 or 2 of this Act is guilty of a crime against future generations.

6. Crime of Ecocide

The right to life is a universal right and where a person, company, organisation, partnership, or any other legal entity causes extensive damage to, destruction of or loss of human and or non-human life of the inhabitants of a territory under sections 1–5 of this Act is guilty of the crime of Ecocide.

7. Crime of Cultural Ecocide

Where the right to cultural life by indigenous communities has been severely diminished by the acts of a person, company, organisation, partnership, or any other legal entity that causes extensive damage to, destruction of or loss of cultural heritage or life of the inhabitants of a territory under sections 1–6 of this Act, is guilty of the crime of cultural Ecocide.

8. Offence of Ecocide

It will be an offence of Ecocide where a person, company, organisation, partnership, or any other legal entity is found to be in breach of section 1 and 7 of this Act.

9. Liability

(a) Any person who pleads guilty or is found guilty of Ecocide under any sections of this Act; or

(b) any person who pleads guilty or is found guilty of aiding and abetting, counselling or procuring the offence of Ecocide, under any sections of this Act shall be liable to be sentenced to a term of imprisonment. Either in addition to or substitution of imprisonment any person convicted of Ecocide can exercise the option of entering into a restorative justice process.

10. Size, Duration, Impact of Ecocide

The test for determining whether Ecocide is established is determined on either one or more of the following factors, which have impact on the severity of diminution of peaceful enjoyment by the inhabitants, namely:-

(a) size of the extensive damage to, destruction of or loss of ecosystem(s);

(b) duration of the extensive damage to, destruction of or loss of ecosystem(s);

(c) impact of the extensive damage to, destruction of or loss of ecosystem(s)

PART II

11. Proceeds of Crime

The provisions of the Proceeds of Crimes Act 2002 will apply in the event of conviction for any offence pursuant to this Act.

Extent

12. Strict Liability

Ecocide is a crime of strict liability committed by natural and fictional persons.

13. Superior Responsibility

(1) Any director, partner, leader and or any other person in a position of superior responsibility is responsible for offences committed by members of staff under his authority, and is

responsible as a result of his authority over such staff, where he fails to take all necessary measures within his power to prevent or to stop all steps that lead to the commission of the crime of Ecocide.

(2) Any member of government, prime minister or minister in a position of superior responsibility is responsible for offences committed by members of staff under his authority, and is responsible as a result of his authority over such staff, where he fails to take all necessary measures within his power to prevent or to stop all steps that lead to the commission of the crime of Ecocide.

(3) With respect to superior and subordinate relationships not described in subsection (1) and (2), a superior is responsible for offences committed by staff under his effective authority, as a result of his failure to exercise authority properly over such staff where he failed to take all necessary measures within his power to prevent or repress their commission or to submit the matter to the competent authorities for investigation.

(4) Any agency purporting to lobby on behalf of (1), (2) or (3) where steps lead to the commission of Ecocide shall be regarded as aiding, abetting, counselling or procuring the commission of the offence.

(5) A person responsible under this section for an offence is regarded as aiding, abetting, counselling or procuring the commission of the offence.

(6) In interpreting and applying the provisions of this section the court shall take into account any relevant judgment or decision of the International Criminal Court.

(7) Nothing in this section shall be read as restricting or excluding:-
    (a) the liability of any superior, or
    (b) the liability of persons other than the superior.

14. Knowledge

(1) Any director, partner or any other person in a position of superior responsibility is responsible for offences committed

by him where his actions result in Ecocide, regardless of his knowledge or intent;

(2) Any member of government, president, prime minister or minister in a position of superior responsibility is responsible for offences committed by him where his actions result in Ecocide, regardless of his knowledge or intent.

15. Withdrawal of immunity of government officials and other superiors

Where any government official and other superior or their members of staff are in breach of Article 2 of the Universal Declaration of Human Rights, after the commencement of this Act, the prosecution may be enforced as of right by proceedings taken for that purpose in accordance with the provisions of this Act.

16. Unlawful use of land

Where any land has been destroyed, damaged or depleted as a result of Ecocide or any offences in this Act, any person who exercises authority over and/or responsibility for the land shall be guilty of that offence and shall be liable to be proceeded against and punished accordingly.

17. Culpability of a company, organisation, partnership, or any other legal entity

(1) Where an offence under any provision of this Act committed by a company, organisation, partnership, or any other legal entity is proved to have been committed with the consent or connivance of, or to have been attributable to any neglect on the part of, any director, manager, secretary or a person who was purporting to act in any such capacity, he as well as the company, organisation, partnership, or any other legal entity shall be guilty of that offence and shall be liable to be proceeded against and punished accordingly.

(2) Where a person of superior responsibility is convicted of an offence under this Act by reason of his position as CEO, director, manager, secretary or a person who was purporting

to act in any such capacity for a company, organisation, partnership, or any other legal entity, as a consequence of the conviction the company shall be held jointly responsible for the actions of its servant.

PART III
Orders
18. Power to order Restoration and Costs
Where any person, company, organisation, partnership, or any other legal entity has committed an offence under this Act:-

(1) a Restoration Order shall be made; and

(2) a Costs Order shall be made; and

(3) the named person, company, organisation, partnership, or any other legal entity that had business in the given territory shall be deemed responsible for the clean-up operations to the extent that the territory be restored to the level it was before the Ecocide occurred.

19. Restorative Justice
(1) Subject to subsection (2), where a defendant pleads or is found guilty, the court must remand the case in order that the victim(s) shall be offered the opportunity to participate in a process of restorative justice involving contact between the offender and any representatives of those affected by the offence.

(2) The court need not remand the case for the purpose specified in subsection (1) where it is of the opinion that the offence was so serious that this would be inappropriate.

(3) The court has the power to order heads of agreement.

(4) Heads of agreement pursuant to a Restorative Justice process can include the following:-
    (i)   Restoration Order
    (ii)   Cost Order
    (iii) Envoronmental Protection Order
    (iv) Suspension of Operations Order
    (v)   Environment Investigation Agency Order

(vi)   Publicity Order
(vii) Enforcement Notice
(viii) Earth Health and Well-being Report

20. Environmental Protection Order (EPO)

Where any person, company, organisation, partnership, or any other legal entity has on the balance of probabilities caused or is likely to cause extensive destruction, damage to or loss of ecosystems of a given territory an EPO shall be made for the duration of any related proceedings and shall only be extinguished by either an acquittal or by an imposition of a Restoration Order.

21. Suspension of Operations Order

Any person, company, organisation, partnership, or any other legal entity identified under a restoration order shall be suspended from operating until the territory has been restored to a level that is acceptable to an independent audit, undertaken by the Environmental Investigation Agency.

22. Determination by the Environmental Investigation Agency

The Environmental Investigation Agency shall determine whether appropriate remediation has been undertaken within the timescale set by the court, and/or whether additional steps (such as the imposition or discharge of an EPO) are necessary, and/or shall identify the nature of remediation outstanding and how best to implement.

23. Publicity Order

Where any person, company, organisation, partnership, or any other legal entity has committed an offence under this Act a Publicity Order may be ordered by the Court setting out:-

(a) the fact of the conviction;
(b) the terms of any restorative justice, remedial and/or commercial prohibition order(s) or any other order the court has made and deems fit for public announcement;
(c) the amount of any financial order;

(d) specified particulars of the offence.

A publicity order can be renewed at any review hearing following a plea of guilty or conviction.

24. Prohibition Notice

(1) Where a person, organisation or government agency can demonstrate on the balance of probabilities that activities that fall within the definition of Ecocide within this Act are at risk of commencing, or have commenced, or are continuing and involve an imminent risk of Ecocide, the court shall issue a Prohibition Notice on the person(s) and/or the company(s) carrying on the process.

(2) Where a person, organisation or government agency can demonstrate on the balance of probabilities that a failure to take steps by any company, organisation, partnership, government department or any other legal entity can lead to an imminent risk of Ecocide, the court shall issue a notice (a 'prohibition notice') on the person(s) and the company(s) carrying on the process.

(3) A Prohibition Notice shall direct that the authorisation shall, until the notice is withdrawn, wholly or to the extent specified in the notice cease to have effect to authorise the carrying on of the process; and where the direction applies to part only of the process it may impose conditions to be observed in carrying on the part which is so authorised.

25. Enforcement Notice

(1) Any person, company, organisation, partnership, or any other legal entity or government agency that is at risk of being prosecuted for Ecocide may be issued with an Enforcement Notice giving an order made by the court to cease all activities that may give rise to Ecocide.

(2) Any person, company, organisation, partnership, or any other legal entity or government agency that has been found guilty of Ecocide shall be issued with an Enforcement Notice giving

an order made by the court to cease all activities that may give rise to Ecocide and pay any consequential losses.

(3) Where an Enforcement Notice has been ordered by a court, an enforcement Notice shall be issued by the Environment Investigation Agency setting out the steps to be taken and specify the period within which those steps must be taken.

26. Earth Health and Well-being Report
Where a territory has been identified as an area at risk of Ecocide or has been named as a territory for the purposes of section 24, an Earth Health and Well-being Report shall be ordered by the court.

27. False written statements tendered in evidence
Where any person tenders a written statement in any proceedings under this Act which he knows to be false or does not believe to be true, he shall be liable to be sentenced to a term of imprisonment.

28. False oral statements tendered in evidence
Where any person tenders evidence in any proceedings under this Act which he knows to be false or does not believe to be true, he shall be liable to be sentenced to a term of imprisonment.

29. Committing Perjury
The Perjury Act 1911 shall have effect as if this Part were contained in that Act.

30. Disclosure of Finances
Any person, company, organisation, partnership, or any other legal entity who is charged with an offence under this Act must provide full disclosure of their finances to the court and failure to disclose by any person ordered by the court for the purposes of this Part shall be liable to be sentenced to a term of imprisonment.

31. Jurisdiction
(1) Where a person commits Ecocide in a different jurisdiction then, notwithstanding that he does so outside England and

Wales, he shall be guilty of committing or attempting to commit the offence against this Act as if he had done so in England or Wales, and he shall accordingly be liable to be prosecuted, tried and punished in England and Wales without proof that the offence was committed there.

(2) Where a person of UK residence is in a different jurisdiction and who is charged with, or found guilty of in absentia, any sections under this Act, a warrant for his arrest shall be issued.

(3) Where there is more than one person, in different jurisdictions and who are charged with, or found guilty of in absentia, any sections under this Act, multiple warrants may be issued at the same time.

Restoration and Consequential Loss Costs

32. Restoration and Consequential Loss Costs

Where any person, company, organisation, partnership, or any other legal entity has been convicted of Ecocide, he and/or it shall be held responsible for any restoration costs that have arisen from causing Ecocide and any consequential losses arising from injury, loss of life, diminution of health or well-being of the inhabitants of the given territory.

33. Balance of Probabilities

No costs shall accrue to any person, organisation or government agency when seeking an order, interim order or prosecution pursuant to the provisions of this Act; costs shall only apply when the person, organisation or government agency fails to establish on the balance of probabilities that there exists a prima facie case pursuant to the provisions of this Act.

34. Costs Assesment

Where Ecocide has occurred, the health and well-being of the community shall be restored as far as possible to the condition as it existed before the Ecocide occurred; and

(1) such costs of cultural Ecocide shall be accorded equal priority with restoration of any ecological Ecocide; and

(2) any costs shall be assessed at a separate cost hearing and shall be enforceable under an Enforcement Notice.

Extent

35. International Criminal Court Act 2001

Section 51 of the International Criminal Court Act 2001, as amended, shall now read:

(1) It is an offence against the law of England and Wales for a person to commit genocide, a crime against humanity and nature, a crime of aggression, a war crime or Ecocide.

(2) This section applies to acts committed:-

      (a) in England or Wales, or

      (b) outside the United Kingdom

by a United Kingdom national, a United Kingdom resident or a person subject to UK service jurisdiction.

36. Short Title, Application and Extent

This Act:-

(1) may be cited as the Ecocide Act 2010;

(2) extends to the whole of the United Kingdom;

(3) may be subject to additions and shall prevail over all other legislation;

No exemptions shall be made subsequent to this Act being enacted.

33. Interpretation

In this Act:-

'Cultural Ecocide' means the damage, destruction to or loss of a community's way of life including a community's spiritual practices.

'Earth Health and Well-being Report' means a report which shall include an assessment of human, cultural and non-human health and well-being impact from damage, destruction to or loss of ecosystem(s) of the immediate and/or any other territories affected or at risk of being affected.

'*ecosystem*' means a biological community of interdependent living organisms and their physical environment.

'*inhabitants*' means any living species dwelling in a particular place.

'*other causes*' means naturally occurring events such as but not limited to; tsunamis, earthquakes, acts of God, floods, hurricanes and volcanoes.

'*peaceful enjoyment*' means the right to peace, health and well-being of all life.

'*restorative justice*' means a process applied as an alternative to conventional sentencing. Where guilt has been accepted or a defendant has been found guilty, he/she may choose to enter into a restorative justice process where he/she shall engage with representatives of parties injured to agree terms of restoration.

'*territory*' means any domain, community or area of land, including the people, water and/or air that is affected by or at risk or possible risk of Ecocide.

ECOCIDE
SENTENCING
GUIDELINES

The following principles emerge as relevant when sentencing in cases of ecocide:-

1. The environment in which we live is a precious heritage, and it is incumbent on the present generation (including the courts) to play a part in preserving it for the future. This may be put more simply as: 'Please leave this planet as you would wish to find it'. [Costing the Earth.]

2. All sentencing for environmental offences and especially the offence of ecocide must strive to promote good environmental governance. It must actively promote effective, participative and collaborative systems of governance at all levels in society – engaging people's creativity, energy, and diversity to ensure that all activity which potentially affects the environment is designed in such a way that the well-being of the planet comes before profit.

3. Sentencing must promote, disseminate and enforce the three fundamental principles which underpin environmental protection:
    (i) The Preventative Principle,
    (ii) The Precautionary Principle, and the
    (iii) The Polluter Pays Principle.

(i)   **The Preventative Principle** requires that the prevention of harm should be the primary aim when taking decisions or implementing action that may have adverse environmental effects. It is consistent with the statutory sentencing purpose of reducing crime. Environmental sentencing may be regarded as having a deterrent effect.

(ii) **The Precautionary Principle** is found in Principle 15 of the Rio Declaration 1992 and provides that where there are threats of serious or irreversible damage, lack of full scientific certainty shall not be used as a reason for postponing cost-effective measures to prevent environmental degradation

(iii) **The Polluter Pays Principle** recognizes the inherent right to life of human and non-human beings. Where pollution has been caused, the polluter has the burden to pay for all costs required, not only to remediate but also to prevent further pollution. Parliament has imposed on CEOs [people in positions of superior responsibility] and companies a heavy burden to do everything possible to ensure that they do not cause pollution.

4. Ecocide is an offence of strict liability precisely because Parliament regards the causing of damage, destruction to or loss of ecosystems on a extensive scale to be so undesirable as to merit the imposition of criminal punishment irrespective of an individual's and/or the company's knowledge, state of mind, belief or intention.

5. The onus is on CEOs and companies to conduct continuing environmental impact assessments looking not only at the likelihood of events occurring that might lead to extensive ecocide, but also at the extent of the damage, or possible damage, if such events do occur. When the level of consequence requires it, fail-safe systems must be put in place.

6. The purposes of sentencing are set out in section 142(1) of the Criminal Justice Act 2003, as amended, in respect of environmental offences as follows:-

'Any court dealing with an offender in respect of an ecocide offence must have regard to the following for purposes of sentencing:–

(d) the making of reparation by offenders to persons affected by their offences;

(e) the protection of people and the planet come first, before profit;

(f) the punishment of the offender who holds a position of superior responsibility;

(g) the reform and rehabilitation of companies and those holding positions of superior responsibility to ensure prevention of ecocide;

(h) the prevention of ecocide by means of deterrence.

Section 143(1) and (2) of the same Act, as amended, provide that:-

(1) In considering the seriousness of any offence, the court must consider the offender's culpability in committing the offence, and the size, extent and duration of the harm that the offence caused, was intended to cause or might foreseeably have caused.

(2) In considering the seriousness of an offence committed by an offender who has one or more previous cautions and/or convictions, the court must treat each previous caution and/or conviction as an aggravating feature if the court considers that it can reasonably be so treated having regard, in particular, to:-

(a) the nature of the offence to which the conviction relates and its   relevance to the current offence,

(b) the time that has elapsed since the conviction.'

7. Punishment, Deterrence And Reparation are all particularly important purposes of sentence in cases of ecocide.

8. Punishment speaks for itself and in conventional terms represents the gravest imposition of incarceration as a public expression and message of deterrence. It further represents an expression of public disquiet that a person in a position of superior responsibility has permitted ecocide on a scale that impinges upon the future well-being of all life.

9. Deterrence. The purpose of deterrence includes:-
(1) making clear that the overall penalty for a breach of the law is always likely to be much more costly than any expense that should have been incurred in avoiding the breach in the first place or that can be passed on to customers as cost outlay;
(2) the need for the overall penalty to be such as to bring the necessary message home to the particular defendant (whether individual and/or corporate) before the Court, in order to deter future breaches – whether by that defendant, or by other potential offenders; and
(3) the need for equal deterrence of all potential offenders, whether wealthy or of limited means – not least because the wealthiest potential offenders are likely, via the scale of their operations, to have the greatest potential to cause the most serious damage.

10. Reparation. The purpose of reparation is to make amends, offer expiation, and make right a wrong or injury. Reparation includes, but is not solely confined to, restorative justice provisions.

11. Seriousness should ordinarily be assessed first by asking:-
(1) How foreseeable was the ecocide? The more foreseeable it was, the graver usually will be the offence.
(2) How far short of the applicable standard did the defendant fall?
(3) How common in this organisation is the kind of breach which led to the environmental pollution/damage? How widespread was the non-compliance? Was it isolated in extent or indicative of a systematic departure from good practice across the defendant's operations?

(4) How far up the organization does the breach go and the degree of culpability within the command structure of the company? Usually, the higher up the responsibility for the breach, the more serious the offence.

12. Sentence of Imprisonment.

(1) If a court concludes the custody threshold has been crossed after taking into account the seriousness factors set out above, then the period of custody is determined by the category into which the convicted person falls, as assessed by his culpability, namely:-

>   (i) Ecocide by dangerous industrial activity – the entry point is 4 years.
>   (ii) Ecocide by reckless knowledge by an objective standard – the entry point 10 years.
>   (iii) Ecocide by intent – the entry point is 12 years or more.

(2) The entry points for custody can be reduced or increased depending on the balance of competing aggravating and mitigating feature examples within a case as set out below.

13. Factors which, if present, are likely to **Aggravate** the offence (the list is not exhaustive):-

(1) loss of human life;

(2) extensive mortality among wildlife;

(3) likely extinction of particular species of wildlife (and/or listed on the International Union for Conservation of Nature Red List of Threatened Species as critically endangered, endangered or vulnerable);

(4) if there is limited prospect of repairing or undoing environmental damage caused;

(5) failure to heed warnings or advice, whether from officials such as the Inspectorate, NGO's and/pressure groups or by employees or other persons, or to respond appropriately to 'near misses' arising in similar circumstances;

(6) cost-cutting at the expense of environmental damage/ pollution. The skimping of proper precautions to make or

save money, or to gain a competitive advantage;

(7)   deliberate failure to obtain or comply with relevant licences, at least where the process of licensing involves some degree of control, assessment or observation by independent authorities;

(8)   a particularly vulnerable environment where wildlife and fauna are affected, especially where a protected species or a site designated for nature conservation was affected;

(9)   Other lawful activities were prevented or significantly interfered with;

(10) Any previous convictions for environmental or environmental related offences.

14. Conversely, the following factors, which are similarly non-exhaustive, are likely, if present, to afford **Mitigation**:-

(1)   a good record of compliance with the law;

(2)   a prompt acceptance of responsibility and timely admission of guilt, and a plea of guilty at an early opportunity;

(3)   a high level of co-operation with the investigation, beyond that which is to be expected;

(4)   genuine efforts to remedy the defect;

(5)   a good environmental awareness and promotion record;

(6)   a responsible attitude to the environment and risks of pollution and damage such as the commissioning of expert advice or the consultation with employees or others affected by the organisation's activities.

(7)   Commission of an ecocide offence may in some cases be established solely by the unauthorized act of an employee. In such a case the responsibility of the organisation and person in position of superior responsibility must be assessed, for example, for inadequate supervision or training. There may be some cases where there is very little culpability in the organisation itself.

(8)   It will generally be appropriate to require the prosecution to set out in writing the facts of the case relied upon and any aggravating or mitigating features which it identifies. The

defence will be required similarly to set out in writing any points on which it differs. If sentence is to proceed upon agreed facts, they should be set out in writing.

(9) In assessing the financial consequences of a fine, the court should consider (inter alia) the following factors:

    (a) the effect on the employment of the innocent may be relevant;

    (b) whether the fine will have the effect of putting the defendant out of business will be relevant; in the worst cases this will be an acceptable consequence.

    (c) the effect on a public organisation such as a local authority or hospital trust; 'The Judge has to consider how any financial penalty will be paid. If a very substantial financial penalty will inhibit the proper performance by a statutory body of the public function that it has been set up to perform, that is not something to be disregarded.' The same considerations will be likely to apply to non-statutory bodies or charities if providing public services.

    (d) the liability to pay civil compensation will ordinarily not be relevant; normally this will be provided by insurance or the resources of the defendant will be large enough to meet it from its own resources;

    (e) the cost of meeting any remedial order will not ordinarily be relevant, except to the overall financial position of the defendant; such an order requires no more than should already have been done.

Any adverse impact upon share price will not be relevant; nor that the prices charged by the defendant company might in consequence be raised.

15. Publicity Orders are to be made at each Sentencing. The object is deterrence and punishment. They may require publication in a specified manner of:

    (a) the fact of conviction;

    (b) the terms of any restorative justice, remedial and/or

commercial prohibit order(s);

(c) the amount of any fine;

(d) specified particulars of the offence.

(1) The order should normally specify the place where public announcement is to be made, and consideration should be given to indicating the size of any notice or advertisement required. It should ordinarily contain a provision designed to ensure that the conviction becomes known to shareholders in the case of companies and local people in the case of public bodies. Consideration should be given to requiring a statement on the defendant's website. A newspaper announcement may be unnecessary if the proceedings are certain to receive news coverage in any event, but if an order requires publication in a newspaper it should specify the paper, the form of announcement to be made and the number of insertions required.

(2) The prosecution should provide the court in advance of the sentencing hearing, and should serve on the defendant, a draft of the form of order suggested and the judge should personally endorse the final form of the order.

(3) Consideration should be given to stipulating in the order that any comment placed by the defendant alongside the required announcement should be separated from it and clearly identified as such.

NEW WORLD BANK ASSESSMENT RULES[1]

WORLD BANK OPERATIONAL POLICY ENVIRONMENTAL
ASSESSMENT

1. The Bank requires environmental assessment (EA) of projects
    proposed for Bank financing to help ensure that they are
    environmentally sound and sustainable, and thus to improve
    decision making.

2. EA is a process whose breadth, depth, and type of analysis
    depend on the nature, scale, and potential environmental
    impact of the proposed project. EA evaluates a project's
    potential environmental risks and impacts in its area of
    influence; examines project alternatives; identifies ways of
    improving project selection, siting, planning, design, and
    implementation by preventing, minimizing, mitigating,
    or compensating for adverse environmental impacts and
    enhancing positive impacts; and includes the process of
    mitigating and managing adverse environmental impacts
    throughout project implementation. The Bank favours
    preventive measures over mitigatory or compensatory
    measures, whenever feasible.

3. EA takes into account the natural environment (air, water, and
    land); human health and safety; social aspects (involuntary
    resettlement, indigenous peoples, and physical cultural

---

1   for original World bank Operational Policy 4.01 Environmental Assessment and Definitions see:
    http://web.worldbank.org/WBSITE/EXTERNAL/TOPICS/ENVIRONMENT/EXTENVASS/0,,cont
    entMDK:20482643~menuPK:1182608~pagePK:148956~piPK:216618~theSitePK:407988,00.html

resources); and transboundary and global environmental aspects.[2] EA considers natural and social aspects in an integrated way. It also takes into account the variations in project and country conditions; the findings of country environmental studies; national environmental action plans; the country's overall policy framework, national legislation, and institutional capabilities related to the environment and social aspects; and obligations of the country, pertaining to project activities, under relevant international environmental treaties and agreements. The Bank does not finance project activities that would contravene such country obligations, as identified during the EA. EA is initiated as early as possible in project processing and is integrated closely with the economic, financial, institutional, social, and technical analyses of a proposed project.

4. The borrower is responsible for carrying out the EA. For Category B projects, the borrower retains independent EA experts not affiliated with the project to carry out the EA. The borrower must engage an advisory panel of independent, internationally recognized environmental specialists to advise on all aspects of the project relevant to the EA. The role of the advisory panel depends on the degree to which project preparation has progressed, and on the extent and quality of any EA work completed, at the time the Bank begins to consider the project.

5. The Bank advises the borrower on the Bank's EA requirements. The Bank reviews the findings and recommendations of the EA to determine whether they provide an adequate basis for processing the project for Bank financing. When the borrower has completed or partially completed EA work prior to the Bank's involvement in a project, the Bank reviews the EA to ensure its consistency with this policy. The Bank will make all EA's available for public consultation and disclosure.

6. The Pollution Prevention and Abatement Handbook describes pollution prevention and abatement measures and emission

---

2   Global environmental issues include climate change, ozone-depleting substances, pollution of international waters, and adverse impacts on biodiversity.

levels that are normally acceptable to the Bank. The handbook meets the standards set by the Ecocide Act. The EA report must provide full and detailed justification for the levels and approaches chosen for the particular project or site.

EA Instruments

7. Depending on the project, a range of instruments can be used to satisfy the Bank's EA requirement: environmental impact assessment (EIA), regional or sectoral EA, strategic environmental and social assessment (SESA), environmental audit, hazard or risk assessment, environmental management plan (EMP) and environmental and social management framework (ESMF). EA applies one or more of these instruments, or elements of them, as appropriate. When the project is likely to have sectoral or regional impacts, sectoral or regional EA is required.

Environmental Screening

8. The Bank undertakes environmental screening of each proposed project to determine the appropriate extent and type of EA. The Bank classifies the proposed project into one of four categories, depending on the type, location, sensitivity, and scale of the project and the nature and magnitude of its potential environmental impacts.

    a. Category A: A proposed project is classified as Category A if it is likely to have significant adverse environmental impacts that are sensitive, diverse, or unprecedented. These impacts may affect an area broader than the sites or facilities subject to physical works. EA for all projects examines the project's potential negative and positive environmental impacts, compares them with those of feasible alternatives (including the 'without project' situation), and recommends any measures needed to prevent, minimize, mitigate, or compensate for adverse impacts and improve environmental performance.

When a project is classified as Category A, a report will be submitted to the relevant authorities for consideration for criminal prosecution.

b. Category B: A proposed project is classified as Category B if its potential adverse environmental impacts on human populations or environmentally important areas – including wetlands, forests, grasslands, and other natural habitats – are less adverse than those of Category A projects. These impacts are site-specific; none are irreversible and mitigatory measures can be designed. The scope of EA for a Category B project may vary from project to project, and will examine the project's potential negative and positive environmental impacts, compares them with those of feasible alternatives (including the 'without project' situation), and recommends any measures needed to prevent, minimize, mitigate, or compensate for adverse impacts and improve environmental performance. The findings and results of Category B EA are described in the project documentation (Project Appraisal Document and Project Information Document).

c. Category C: A proposed project is classified as Category C if it is likely to have minimal or no adverse environmental impacts. Beyond screening, no further EA action is required for a Category C project.

d. Category FI: A proposed project is classified as Category FI if it involves investment of Bank funds through a financial intermediary, in subprojects that may result in adverse environmental impacts.

EA FOR SPECIAL PROJECT TYPES
Sector Investment Lending

9. For sector investment loans (SILs), during the preparation of each proposed subproject, the project coordinating entity or implementing institution carries out appropriate EA

according to country requirements and the requirements of this policy. The Bank appraises and, if necessary, includes in the SIL components to strengthen, the capabilities of the coordinating entity or the implementing institution to:-

  a. screen subprojects,
  b. obtain the necessary expertise to carry out EA,
  c. review all findings and results of EA for individual subprojects,
  d. ensure implementation of mitigation measures (including, where applicable, an EMP), and
  e. monitor environmental conditions during project implementation.

If the Bank is not satisfied that adequate capacity exists for carrying out EA, all Category A subprojects and, as appropriate, Category B subprojects – including any EA reports – will be refused.

Financial Intermediary Lending

10. For a financial intermediary (FI) operation, the Bank requires that each FI screen proposed subprojects and ensure that subborrowers carry out World Bank standard of EA for each subproject. Before approving a subproject, the FI must verify (through its own staff, outside experts, or existing environmental institutions) that the subproject meets the environmental requirements of the Ecocide Act and is consistent with this OP.

11. In appraising a proposed FI operation, the Bank reviews the adequacy of country environmental requirements relevant to the project and the proposed EA arrangements for subprojects, including the mechanisms and responsibilities for environmental screening and review of EA results. When necessary, the Bank ensures that the project includes components to strengthen such EA arrangements. For FI operations expected to have Category A subprojects, prior to the Bank's appraisal each identified participating FI provides to the Bank a written assessment of the institutional mechanisms

(including, as necessary, identification of measures to strengthen capacity) for its subproject EA work. Where the Bank is not satisfied that adequate capacity exists for carrying out EA, all subprojects – including EA reports – are subject to prior review and approval by the Bank and will be submitted for consideration for criminal prosecution.

Emergency Operations under OP 8.00

12. The policy set out in OP 4.01 normally applies to emergency operations processed under OP/BP 8.00, Rapid Response to Crises and Emergencies. However, when compliance with any requirement of this policy would prevent the effective and timely achievement of the objectives of an emergency operation, the Bank may exempt the project from such a requirement. The justification for any such exemption is recorded in the loan documents. In all cases, however, the Bank requires at a minimum that (a) the extent to which the emergency was precipitated or exacerbated by inappropriate environmental practices be determined as part of the preparation of such projects, and (b) any necessary corrective measures be built into either the emergency operation or a future lending operation.

Institutional Capacity

13. When the borrower has inadequate legal or technical capacity to carry out key EA-related functions (such as review of EA, environmental monitoring, inspections, or management of mitigatory measures) for a proposed project, the project includes components to strengthen that capacity.

Public Consultation

14. For all Category B projects proposed for IBRD or IDA financing, during the EA process, the borrower consults project-affected groups and local nongovernmental organisations (NGOs) about the project's environmental aspects and takes their views and must justify any departure

from their recommendations in the final decision. The borrower initiates such consultations as early as possible. For Category B projects, the borrower consults these groups at least twice: (a) shortly after environmental screening and before the terms of reference for the EA are finalized; and (b) once a draft EA report is prepared. In addition, the borrower consults with such groups throughout project implementation as necessary to address EA-related issues that affect them.

Disclosure

15. For meaningful consultations between the borrower and project-affected groups and local NGOs on all projects proposed for IBRD or IDA financing, the borrower provides relevant material in a timely manner prior to consultation and in a form and language that are understandable and accessible to the groups being consulted.

16. For a Category B project, the borrower provides for the initial consultation a summary of the proposed project's objectives, description, and potential impacts; for consultation after the draft EA report is prepared, the borrower provides a summary of the EA's conclusions. In addition, for a Category B project, the borrower makes the draft EA report available at a public place accessible to project-affected groups and local NGOs. For SILs and FI operations, the borrower/FI ensures that EA reports for all subprojects are made available in a public place accessible to affected groups and local NGOs.

17. Any separate Category B report for a project proposed for IDA financing is made available to project-affected groups and local NGOs. Public availability in the borrowing country and official receipt by the Bank of Category B reports for projects proposed for IBRD or IDA financing are prerequisites to Bank appraisal of these projects.

18. Once the borrower officially transmits the Category B EA report to the Bank, the Bank distributes the summary (in English) to the executive directors (EDs) and makes the report available through its InfoShop. If the borrower objects

to the Bank's releasing an EA report through the World Bank InfoShop, Bank staff (a) do not continue processing an IDA project, or (b) for an IBRD project, submit the issue of further processing to the EDs.

Implementation

19. During project implementation, the borrower reports on (a) compliance with measures agreed with the Bank on the basis of the findings and results of the EA, including implementation of any EMP, as set out in the project documents; (b) the status of mitigatory measures; and (c) the findings of monitoring programs. The Bank bases supervision of the project's environmental aspects on the findings and recommendations of the EA, including measures set out in the legal agreements, any EMP, and other project documents.

Definitions

1. *Environmental audit*: An instrument to determine the nature and extent of all environmental areas of concern at an existing facility. The audit identifies and justifies appropriate measures to mitigate the areas of concern, estimates the cost of the measures, and recommends a schedule for implementing them. For certain projects, the EA report may consist of an environmental audit alone; in other cases, the audit is part of the EA documentation.

2. *Environmental impact assessment (EIA)*: An instrument to identify and assess the potential environmental impacts of a proposed project, evaluate alternatives, and design appropriate mitigation, management, and monitoring measures. Projects and subprojects need EIA to address important issues not covered by any applicable regional or sectoral EA.

3. *Environmental management plan (EMP)*: An instrument that details (a) the measures to be taken during the implementation and operation of a project to eliminate adverse environmental impacts; and (b) the actions needed to implement these measures. The EMP is an integral part of Category B EAs

(irrespective of other instruments used). EAs for Category C projects may also result in an EMP.

4. *Environmental and social management framework (ESMF)*: An instrument that examines the issues and impacts associated when a project consists of a program and/or series of sub-projects, and the impacts cannot be determined until the program or sub-project details have been identified. The ESMF sets out the principles, rules, guidelines and procedures to assess the environmental and social impacts. It contains measures and plans to reduce, mitigate and/or offset adverse impacts and enhance positive impacts, provisions for estimating and budgeting the costs of such measures, and information on the agency or agencies responsible for addressing project impacts. The term 'Environmental Management Framework' or 'EMF' may also be used.

5. *Hazard assessment*: An instrument for identifying, analyzing, and controlling hazards associated with the presence of dangerous materials and conditions at a project site. The Bank requires a hazard assessment for projects involving certain inflammable, explosive, reactive, and toxic materials when they are present at a site in quantities above a specified threshold level. For certain projects, the EA report may consist of the hazard assessment alone; in other cases, the hazard assessment is part of the EA documentation.

6. *Project area of influence*: The area likely to be affected by the project, including all its ancillary aspects, such as power transmission corridors, pipelines, canals, tunnels, relocation and access roads, borrow and disposal areas, and construction camps, as well as unplanned developments induced by the project (e.g., spontaneous settlement, logging, or shifting agriculture along access roads). The area of influence may include, for example,

    a.   the watershed within which the project is located;

    b.   any affected estuary and coastal zone;

    c.   off-site areas required for resettlement or compensatory tracts;

d.   the airshed (e.g., where airborne pollution such as smoke or dust may enter or leave the area of influence;

e.   migratory routes of humans, wildlife, or fish, particularly where they relate to public health, economic activities, or environmental conservation; and

f.   areas used for livelihood activities (hunting, fishing, grazing, gathering, agriculture, etc.) or religious or ceremonial purposes of a customary nature.

7.   *Regional EA*: An instrument that examines environmental issues and impacts associated with a particular strategy, policy, plan, or program, or with a series of projects for a particular region (e.g., an urban area, a watershed, or a coastal zone); evaluates and compares the impacts against those of alternative options; assesses legal and institutional aspects relevant to the issues and impacts; and recommends broad measures to strengthen environmental management in the region. Regional EA pays particular attention to potential cumulative impacts of multiple activities.

8.   *Risk assessment*: An instrument for estimating the probability of harm occurring from the presence of dangerous conditions or materials at a project site. Risk represents the likelihood and significance of a potential hazard being realized; therefore, a hazard assessment often precedes a risk assessment, or the two are conducted as one exercise. Risk assessment is a flexible method of analysis, a systematic approach to organizing and analyzing scientific information about potentially hazardous activities or about substances that might pose risks under specified conditions The Bank routinely requires risk assessment for projects involving handling, storage, or disposal of hazardous materials and waste, the construction of dams, or major construction works in locations vulnerable to seismic activity or other potentially damaging natural events. For certain projects, the EA report may consist of the risk assessment alone; in other cases, the risk assessment is part of the EA documentation.

9. *Sectoral EA*: An instrument that examines environmental issues and impacts associated with a particular strategy, policy, plan, or program, or with a series of projects for a specific sector (e.g., power, transport, or agriculture); evaluates and compares the impacts against those of alternative options; assesses legal and institutional aspects relevant to the issues and impacts; and recommends broad measures to strengthen environmental management in the sector. Sectoral EA pays particular attention to potential cumulative impacts of multiple activities.

10. *Strategic environmental and social assessment (SESA)*: An instrument that describes analytical and participatory approaches that aim to integrate environmental and social considerations into policies, plans and programs and evaluate their inter linkages with economic considerations. The term 'Strategic Environmental Assessment' or 'SEA' may also be used.

FREQUENTLY ASKED QUESTIONS AND ANSWERS

What are the two types of ecocide?

Human-made ecocide and naturally occurring ecocide. Human made ecocide includes the loss of the Amazon, mining, the Athabasca Tar Sands in Canada and a nuclear war.

Naturally occurring ecocide include rising sea levels, tsunamis, floods and earthquakes. Human ecocides can be prevented, naturally occurring cannot. By creating a Law of Ecocide, business, banks and nations will be under a legal duty of care to ensure that profit, money and policy does not support mass damage and destruction of the Earth by humanity.

What will a Law of Ecocide do?
- By legally defining ecocide, a legal duty of care is created. Companies will require a period of transition whereby no prosecutions are pursued whilst they change their practices from 'polluter pays' to 'polluter doesn't pollute'.
- Finance of dangerous industrial activity will be withdrawn. Bridging loans will be required during the transition period to assist companies withdraw from certain illegal activities without threat of prosecution.
- Nature and humanity will be prioritised. Inhabitants, both human and non-human, will accrue the legal right to peace.

- By placing ecocide on the same legal footing as the international crime of genocide, a superior law which overrides national laws will create a global level playing-field for all.
- Plans can be made to discharge the legal duty of care on all nations to provide assistance to those territories at risk of ecocide in advance.
- A Law of Ecocide will stop the flow of destruction at source. By going upstream to the source of the problem (where the ecocide occurs), it is much cheaper to prevent it the first place. This is good for economies, people and planet. It is always far more expensive to remedy something after it has happened. A Law of Ecocide is preventative, pre-emptive and post-operative.

Is there business support for the amendment to the Rome Statute?
Yes there is. A Law of Ecocide provides the missing legislative framework to enable private and public capital to flow into emerging technology providers and to impose on governments the legal duty to drive environmental improvements as the primary economic growth strategy. By enabling the transition to the green economy, business can plan ahead and investors will have security of long-term indicators and future market trends. It will accelerate gigaton-scale solutions to climate change.

Who would initiate the legal proceeding of ecocide?
As the International Criminal Court (ICC) is a court of last resort, the starting point is the state. Where a Member state is either unwilling or unable to act, the ICC will step in.

If, for example, it is in relation to activities licensed by a government, eg. the Alberta tar sands?
A case can be raised by the ICC in any one of 4 ways; by the UN, by a Member state, by a prosecutor of the ICC of his own doing or by an individual writing to the ICC.

In the example of the Alberta Tar Sands, in Canada individuals can raise private prosecutions. Where the State fails to act, the ICC can step in.

How will governments such as Canada be bound by a law of Ecocide?

Canada has already ratified the Rome Statute. What is being sought is an amendment to include a 5th Crime under the Rome Statute. Once it is amended, Canada will be bound by it (see below for further details of amendment process).

Who will enforce the judgment of the ICC? What are the penalties and how will they work?

Enforcement of any ICC conviction is by a sentence of imprisonment. All international crimes begin with a minimum of two years.

Has anything yet been written on this, particularly in the way of critical scholarly work.

Many academic institutions are using the book, *Eradicating Ecocide*, as a course book – law schools, environmental studies and business schools. Students are writing theses and papers are being written for publication, some later this year. In May 2012, Oslo University held a conference on the Rule of law for Nature; they have received over 100 submissions from over fifty countries.

Whose interests are being protected?

By creating a Law of Ecocide, the law will protect people and planet first.

How will a Law of Ecocide affect our engagement with nature?

Stewardship will become the number one priority superceding ownership, which will become a secondary priority to the primary duty of care owed to others.

Why does the Law of Ecocide include activities carried out by human agency as well as by other causes?

By including naturally occurring ecocide a legal duty of care is created to provide assistance to those who are at risk of mass ecosystem collapse, whether it be as a result of rising-sea levels or a tsunami.

What is the legal remedy for a naturally occurring event that causes environmental devastation?
Firstly, nations will be held legally accountable for helping those who have been, or are at risk of, subjected to naturally occurring ecocide. Secondly, in so doing, emergency relief will become a legal requirement.

Which company director is charged?
It could be the CEO and/or any one of or all of the directors (and any head of bank who has financed the dangerous industrial activity). The decision as to whom to prosecute will ultimately lie with the prosecution.

What is the scale of the environmental damage done to be a crime of ecocide?
This comes down to size, duration and impact. During war-time, the ENMOD Convention sets outs parameters; the same can be applied during peace-time.

Is each director of each company charged or a collective of companies within an industry?
This is a decision for a prosecutor; whether to use sample counts against sample industries or to take a belt and braces approach and prosecute them all.

Who would be charged for damage in relation to activities, for example in the Amazon – the farmer or the government?
Depending on the size, duration and impact test, it can be the owner, a minister and/or the bank who financed it.

Will criminalisation of this kind of corporate activity shift business practices in reality?
Yes. Ecocide is already a crime in Kyrgyzstan and currently a prosecution is underway.

What is the change model?

A transition period of five years.

What are the practical workings of a crime of ecocide?

Once a law of ecocide is made international law, nations will pass laws to include ecocide as a crime. The crime will then be part of national law. In September 2010 a mock trial was held at the UK Supreme Court as if this had already happened. Two legal teams tested the law of ecocide. The event was live-streamed across the world and the outcome was two convictions.

How is restorative justice enforced?

It is offered as an alternative sentencing option; parties can attend a restorative justice hearing to come to some form of agreement with parties that have been adversely affected by the ecocide.

What are the pre-conditions for restorative justice to be used/ not used?

Companies, banks and governments accept responsibility for restoring territories adversely impacted. Parties agree to meet with each other and with representatives of the parties who have harmed, to engage in a process where all parties can speak. The aim of the process is to transform and restore; it is solution based and non-blame driven.

Restorative justice is already used within the criminal justice process in many countries. Expanding the remit to include individuals who are in a position of superior responsibility has the power to resolve seemingly standstill disputes.

Restorative justice was introduced as an alternative to conviction at a Restorative Justice Hearing which was held on 31 March 2012. University of Essex hosted a mock post-trial sentence where the principles were explored. Top experts road-tested the process. The event was live-streamed and was open to the public. You can read more here: www.idcr.org.uk/ecocide-trial-the-sentence

Who decides whether a restorative justice sentencing is used?
It is an option for defendants. It can also be used out-with a court process by consenting parties. Sentencing guidelines will be published after the event.

Are they exempt from going to prison?
This can depend on a number of factors when it comes to sentencing; intent, knowledge, mitigation and whether the Restorative Justice hearing has been successful.

How effective is the UN? Will this stop ecocide?
The effectiveness is the power that is vested in the UN to criminalise mass damage and destruction, just as we criminalised genocide. Yes, we still have genocide (and yes we still have smaller crimes, eg theft), however the important issue is that what was once the norm becomes the exception. At the moment we accept mass damage and destruction of the earth because there is no criminal law against it in peace time. When we criminalise the mass destruction of the earth it becomes the exception instead of the norm under which we will operate and most importantly, a legal framework is established within which to seek justice and remedy (not fines, instead the options of imprisonment and/or restorative justice).

Can the Security Council veto the amendment?
No. A member of the Security Council cannot veto a crime when it is established. The legal term is called 'ergo omnes' – when ecocide is made an international crime it applies to all. (note: the Security Council can veto a legal opinion handed down by the International Court of Justice, but the ICJ is not the body that will house the international crime of ecocide – it will be the International Criminal Court or a newly established International Court of the Environment).

For more information please see
www.eradicatingecocide.com

INDEX

ABOUT THE AUTHOR

Polly Higgins is an international lawyer, barrister and advocate for a new body of law, Earth Law. She was the initiator of the idea that humans have a legal duty of care to the earth and non-human beings. Out of that initial idea came proposals to the UN for Earth rights and the Law of Ecocide which she has defined as the 'extensive damage, destruction to or loss of ecosystems of a given territory, whether by human agency or by other causes, to such an extent that peaceful enjoyment by the inhabitants of that territory has been severely diminished.'

Polly studied law and practiced as a barrister in London. In 2008 Polly was invited to speak at a UN conference on her proposal of rights for the earth. She proposed a Universal Declaration, akin to the Universal Declaration of Human Rights. The Declaration has been developed and named the Universal Declaration of the Rights of Mother Earth by Bolivia. The Declaration was the beginning of a process of building laws to protect nature and future generations. Earth rights and ecocide are two sides of a coin.

In 2010 Polly submitted a proposal to the UN setting out the law to ensure the Earth's right to life. Her proposal for a Law of Ecocide was published in her first book, *Eradicating Ecocide: Laws and Governance to Prevent the Destruction of our Planet*, which won The People's Book Prize for non-fiction in 2011. In September 2011 a mock ecocide trial was staged in the UK Supreme Court.

Winner of Campaigner of the Year and Overall Champion at the PEA Awards in 2012, Polly has received various nominations, including the Planet's Lawyer (2010 Performance Awards), she has been named as 'One of the World's Top 10 Visionary Thinkers' by *The Ecologist* and hailed by *The Guardian* as 'one of their Green Heroes for working for the right kind of environmental change'. Polly has spent the past four years advising and speaking on Earth Law. She is an international speaker and is a lawyer for the earth.

Excellent read.

Excellent job Polly! Best read in 2011 so far!

Simply excellent.

Excellent!

Inspirational.

This is a major new approach towards understanding what we are doing to the planet. Corporate business – and others in society – only seem to understand the 'law' and act accordingly to minimise their obligations. Hence we have labour laws, and social laws (although these are often under threat), so to introduce the the term of 'ecocide' is a fantastic development and one we should all welcome and the book has to be considered as a mainstream issue and to win the People's Book Prize would give this more consideration. Thanks Polly.

Too late to vote (by a few minutes) but I must comment to say that this is by far the most important and world-changing book on the list. It is factual and careful, but also very moving. Making ecocide an international crime against peace would go so far towards arresting the destruction and pollution of our only planet.

Best book 2011 for our readers!

Best read 2011 for me.

Met Lawyer loves this book!

Not only a brilliant book but a guide! Thank you Polly.

Important stuff.

Probably the most important book in all history …

Hurrah for Polly.

Ecocide as a UN amendment could transform the face of the planet – as did the abolition of slavery in the C19.

I totally support and agree with Polly's goal for international legislation against ecocide and hope that winning this award will stir more public support.

I vote for Polly Higgins' *Eradicating Ecocide*.

This is an important, inspirational and visionary book which should be read by all world leaders!

Polly gets my vote. Our planet needs to be heard.

Excellent read.

No more deserving and inspirational book and author could win this prize.

Enlightening and frightening – and essential reading. We need to think differently and faster.

This book needs to get out there.

Nam-myoho-renge-kyo.

Polly Higgins is an inspiration. An international lawyer, she has focused her talents towards the creation of a new international law, Ecocide. Higgins is petitioning the UN to add a Fifth Crime Against Peace, following Genocide, War Crimes, Crimes Against Humanity and Crimes of Aggression. If successful this would allow for the prosecution of major polluters at the International Criminal Court in the Hague. This includes Heads of State as well as CEOs. This will be an incredibly difficult task but, having met her and realised the depth of her research and extent of her networking in Southern nations, if anyone can pull it off Higgins can. The next couple of years might be very interesting.

This book should be compulsory reading for all people, businesss and change-makers to inspire us all to lobby our leaders to ACT now and much much faster than they are.

This is the most serious issue facing us; we must not deny it and Polly Higgins helps us face reality.

We need to stop the destruction of natural habitat around the world and the current anthropogenic mass extinction. The proposals in this book will help in this goal.

This book will help create a new foundation for a new humanity!

A book of universal influence and initiative that will stimulate the changes the planet needs.

Fabulous user-friendly legal book which manages to be easy and interesting to read, while being packed full of information. It's also a refreshing positive climate change book.

Making ecocide a crime is SO important that everything else pales into insignificance. If the human race continues to wreak havoc on the Earth's ecosystems there will soon be no more books, no more people, no more life as we know it.

Best book this decade.

It's about time it was made a serious crime to cause Ecocide, and this book says it all. Fantastic!

Looks Great!!

An amazing idea! I hope this book leads to a healthier future for generations to come. I vote for a scientific approach to responsible management of the Earth's resources and for the outcome of such studies to write the rule book for what companies can and cannot do when impacting on the Earth in any way.

WOOT!

Brilliantly written.

This is an important book and important legislation! The Earth is a Living and Intelligent Being, as we are. If we don't stop raping her and killing her, bit by bit; there will be no place for us to live.

Great cause!

I guess the survival life on planet Earth is a concept to get behind, eh?

A heroic and visionary book!

Maybe the most important book for the 21st century!

An important book!

An inspiring read to shake us out of complacency. A call to a new practical way of saving our trees and our planet at the 11th hour.

This book proposes a method to bring to justice those who create wanton destruction to our environment and who, up to now, have been able to hide behind a cloak of corporate invisibility. It removes a huge roadblock to environmental sustainability. Bravo!

A planet-saving read!

Polly Higgins pioneering visionary work is inspiring and this book is essential reading for us all to act on; this is just the start!

A significant book for the new era. Gives hope to all that care about the future. Will inform the environmental activism of millions. Well written and presented.

Eradicating Ecocide sets out how we can 'pull up the tracks' on our unjust legal system and put in place radical new laws which will criminalise behaviours which have a deadening effect on life. Polly Higgins is an activist lawyer and a true inspiration to all young lawyers.

This book is wonderful and deserves to win!

Most needed book in these times … Thanks Polly.

Recyle, renew and reuse the laws that are in place. Eco-LOGICal.

Fantastic that this book is being considered, brilliant.

Great book a must read.

Essential reading. Ecocide must be eradicated or it will destroy us all.

The simple facts and the simple solution to guarantee a sustainable future for our children – a true call to action for every one of us.

A book for our times and a chance for real quality of life.

A must-read for anyone looking for a real and practical solution to the global environmental and cultural crisis.

Surely, this accessible yet comprehensive book counts among the most timely currently available on crucial theme of our times.

A great contribution to a critically important subject area.

Beyond important – we cannot carry on as if there is no tomorrow, as there will be no tomorrow.

This is a major contribution to high-efficacy strategy; to ignore it could be interpreted as a form of sui-genocide.

Polly knows law inside out. She also understands what matters in life. When these two come together, you have an explosive combination generating invaluable insights inspiring for action that counts.

Quite simply the best book on the most important topic of all time. Of course I recommend it. :o)

In you own words Polly 'Trim tabs each and every one of us.' We can all trim direction with our purchases or investments. When CEOs can be prosecuted, they will have to either change direction ... or suffer a personal loss (rather than shareholders 'picking up the Tab!') The Global Law of Ecocide will trigger that change of direction.

Polly Higgins is woman with a huge heart and great vision. She's waking up people all over the world to the need to protect our planet, and hence ourselves. She's right – we have to change the rules so that governments and corporations can no longer see ecocide as an option.

The nuclear tests in the 40's, 50's and 60's and so on were deliberate and ignorant. We cannot learn, despite this wonderful book.

This is an important book. The simple fact of defining Ecocide as a crime could give pause to thinking members of some companies' boards, and hold back destructive projects.

Truly inspiring – how laws could save the planet. This is insightful and proactive – it is at the forefront of a new wave of thought that offers real solutions to tackling global problems.

www.peoplesbooksprize.com

The Predator Culture
The Systemic Roots and Intent of Organized Violence
BY FRED HARRISON

'You can become wealthy by creating wealth or by appropriating the wealth created by other people. When the appropriation of the wealth is illegal it is called theft or fraud. When it is legal economists call it rent-seeking'   JOHN KAY, FINANCIAL TIMES

Fred Harrison draws on global-wide case studies to show how the violent birth of nation-states, whether the result of territorial conquests or colonialism, splits the population into two classes, victors and vanquished. This division is perpetuated and legitimated through the system of land tenure. The pathological consequences – as diverse as failed states, organized crime (Mafia), religious fundamentalism and the re-emergence of piracy – are the result of the violent uprooting of the original inhabitants from their homelands.

Understanding the territorial basis of political power and wealth is the pre-requisite, Fred Harrison argues, for making sense of issues as diverse as genocide, narco-gangsterism, terrorism and fascism. The struggle over land and resources, he contends, is at the root of all of today's global crises. Some attempts are being made to restore land to those in need, ranging from the offer of land in Afghanistan to the Taliban as an inducement to set aside their violent strategies, to the sharing of the rents of oil in Nigeria to entice eco-warriors into mainstream politics. But these piecemeal tactics fail to synthesize the conditions for peace and prosperity.

The Predator Culture provides a framework for truth and reconciliation in what has become a violent world that is slipping dangerously out of control.

Bibliography • Index • Paperback, 192pp, £17.95
For more books on Ethical Economics visit
www.shepheard-walwyn.co.uk • www.ethicaleconomics.org.uk